Unified Theory of the Mechanical Behavior of Matter

UNIFIED THEORY OF THE MECHANICAL BEHAVIOR OF MATTER

M. J. Marcinkowski

Department of Mechanical Engineering, and
Engineering Materials Group
University of Maryland
College Park, Maryland

The work reported in this book was supported by the Department of Energy (Division of Materials Science, Office of Basic Energy Sciences) under Contract EY-76-S-05-3935.

A Wiley-Interscience Publication

JOHN WILEY AND SONS
New York • Chichester • Brisbane • Toronto

Library of Congress Cataloging in Publication Data

Marcinkowski, M J

Unified theory of the mechanical behavior of matter.

"A Wiley-Interscience publication."

Includes index.

1. Physical metallurgy. 2. Strength of materials.
I. Title.

TN690.M2648 669′.9 78-27799

ISBN 0-471-05434-8

Printed in the United States of America

10 9 8 7 6 5 4 3 2 1

To my children Jeffrey and Susan
without whose inspiration
this effort would have been
made much more difficult
if not impossible.

Preface

This book had its origin at a seminar I gave at the Max Planck Institut für Metallforschung in Stuttgart, in the Federal Republic of Germany, on September 10, 1973. After extensive discussions with Professors E. Kröner, A. Seeger, and K. -H. Anthony, it became clear that very serious deficiencies existed with respect to the fundamental meaning of dislocations, disclinations, grain boundaries, and to the mechanical behavior of matter in general. As a result of the mutual stimulation generated by these discussions, I was granted a Senior U.S. Scientist Award presented by the Alexander von Humboldt Stiftung in conjunction with a one-year sabbatical leave from the University of Maryland during 1975. In particular, I was invited to pursue these problems in Professor E. Kröner's Institut für Theoretische und Angewandte Physik der Universität Stuttgart in the Federal Republic of Germany. It was decided that the most general mathematical methods with which to pursue this goal were those of differential geometry with its almost exclusive reliance on tensor calculus. The approach therefore embodies a continuum description of matter.

One of the aims of this book is to show that when the continuum approach is employed rigorously, nearly all of the problems dealing with the mechanical behavior of solids can be solved with satisfactory accuracy. The degree of accuracy is particularly noteworthy with respect to the description of high-angle boundaries where it has frequently been felt that the only approach to this problem should be in terms of the atomic structure. The difficulty here, however, is that such problems are too complex to be formulated exactly, that is, quantum mechanically. It seems, therefore, that an exact continuum description of such problems is more desirable than an approximate atomistic picture. In the former case, the generation of physical concepts is shown to be much more extensive and rewarding, especially as concerns the precise definition of a plastic versus an elastic distortion that includes the concepts of plastic and elastic rotations. The treatment of all these problems has been for the most part nonlinear and thus applicable to generalized distortions. Perhaps the most powerful concept to emerge from the present study is that of a surface dislocation that allows any boundary value problem to be solved either analytically or numerically with the aid of a digital computer. In particular, it is possible to use such surface dislocation arrays to describe any state of stress, either internal or external or a combination of both.

It is difficult for me to express in words my sincere thanks and appreciation to Professor Kröner for the kindness and inspiration afforded me while in Stuttgart. I am also indebted to Professor L. E. Popov and his colleagues of the Institute of Civil Engineers, Department of Physics, Tomsk, U.S.S.R., for their kind invitation to present many of the items contained in the present monograph in a series of mutually stimulating lectures in Novosibirsk from September 12–18, 1976. I would also like to express my appreciation to Dr. R. de Wit of the Metallurgy Division and Institute for Materials Research of the National Bureau of Standards, Washington, D.C., and to Professor R. W. Armstrong of the Department of Mechanical Engineering, University of Maryland, College Park, Maryland for numerous conceptual discussions dealing with the mechanical behavior of solids. Particular gratitude is extended to Dr. K. Jagannadham whose unbounded energy and enthusiasm contributed to the discovery and elucidation of many new concepts in this important area of science. Special thanks and appreciation is also extended to Ms. Dorothea F. Brosius whose pleasant perseverance and warm personality enabled her to convert my frequently unintelligible handwriting into a typewritten work of art. Support for the present effort has and is presently being provided by the Materials Science Program, Division of Physical Research of the U.S. Department of Energy.

Finally, I would like to point out that the present monograph is not intended to be a review of the mechanical behavior of matter but rather represents my own "world view" of the subject, a view that I find to be successful in solving problems in this area. In this respect, I would like to express my apologies for the possible omission of any number of worthy and relevant works carried out by others in this field.

M. J. MARCINKOWSKI

College Park, Maryland
March 1979

Contents

List of Formulae and Symbols

(K)	State associated with the perfect reference crystal, that is, uppercase Latin letters.
$\underset{K}{ds}$	Element of length in the (K) state.
a_{KL}	Metric tensor associated with the (K) state.
dx^K	Coordinate in the (K) state.
$\mathbf{e}_K$	Base vector in the (K) state.
$(\underset{K}{ds})^2 = a_{KL}\, dx^K dx^L$	Definition of length in the (K) state.
$a_{KL} = e_K \cdot e_L$	Definition of metric tensor for the (K) state.
δ_{KL}	Kronecker delta.
(κ)	State associated with the elastically deformed crystal, that is, lowercase Greek letters.
B_κ^K	Elastic distortion tensor.
$\mathbf{e}_\kappa = B_\kappa^K \mathbf{e}_K$	Relation between base vectors in two different states.
$A_K^\kappa = \delta_K^\kappa$	Condition for dragging of coordinates.
$dx^\kappa = A_K^\kappa dx^K$	Relation between coordinates in two different states.
$g_{\kappa\lambda}$	Metric tensor for the (κ) state.
$\mathbf{e}_{\kappa\lambda} = \frac{1}{2}(g_{\kappa\lambda} - a_{\kappa\lambda})$	Elastic strain tensor associated with going from state (K) to state (κ) expressed in terms of dragged coordinates.
$V_{(\kappa)} = \det(B_\kappa^K)$	Volume associated with the (κ) state.
$g_{\kappa\lambda} x^\lambda = \lambda a_{\kappa\lambda} x^\lambda$	Condition for finding principal direction.
C_κ^K	Coordinate transformation.
A_K^a	Plastic distortion tensor.
$B_K^a = \delta_K^a$	Condition for dragging of the metric tensor.
$\mathbf{e}_{KL}^P = \frac{1}{2}(a_{ab} A_K^a A_L^b - a_{KL})$	Plastic strain tensor associated with going from state (K) to state (a) expressed in terms of a dragged metric tensor.

(a)	State associated with a pure plastic distortion, that is, lowercase Latin letters towards front of alphabet.
(k)	States comprised of single dislocations. Denoted by lowercase Latin letters that occupy center of alphabet.
(k')	States comprised of planar dislocation arrays. Denoted by primed lowercase Latin letters that occupy center of alphabet.
$H(+x^1)$	Heaviside function defined as 1 for $x^1>0$ and 0 for $x^1<0$.
$\delta(x^2)$	Dirac delta function defined as 0 for $x^2\neq 0$.
$\int_{-\infty}^{+\infty}\delta(x^2)dx^2=1$	Relation satisfied by the Dirac delta function.
$b^k=-\oint A_K^k dx^K$	Line integral that gives the closure failure associated with a given Burgers circuit.
$\boldsymbol{b}=b^k\boldsymbol{e}_k$	Invariant form of the Burgers vector in terms of its components.
$\oint A_K^k dx^K=\int_s \partial_{[L}A_{K]}^k dF^{LK}$	Conversion of a line integral into a surface integral by means of Stokes' theorem.
$S_{lm}^{\cdot\cdot k}=A_l^L A_m^K \partial_{[L}A_{K]}^k$	Expression for the torsion tensor $S_{lm}^{\cdot\cdot k}$.
$b^k=-\int_s S_{lm}^{\cdot\cdot k}dF^{lm}$	Surface integral that gives the closure failure associated with the dislocation content of a given Burgers circuit.
$\alpha^{nk}=-\varepsilon^{nlm}S_{lm}^{\cdot\cdot k}$	Expression for the dislocation density tensor α^{nk}.
$\varepsilon^{nlm}=\dfrac{e^{nlm}}{\sqrt{a}}$	Expression for the permutation tensor ε^{nlm}.
e^{nlm}	Permutation symbol that is equal to $+1$ or -1 depending upon whether $n,l,m=1,2,3$ is an even or odd permutation, respectively, of this sequence. It is equal to zero if any two or three of the superscripts are equal.
$(a^c),(K^c),(k'^c)$	Representation of the $(a),(K)$, and (k') states, respectively, in terms of a new coordinate system.
Γ_{ml}^k	Coefficient of connection.
$\left\{ {\kappa \atop \lambda\mu} \right\}$	Christoffel symbol of the second kind.
$dC^k=-\Gamma_{ml}^k C^m dx^l$	Change in length for vector C^m upon parallel transport through a distance dx^l.
$\Omega_{cb}^{\cdot\cdot a}=C_c^m C_b^l \partial_{[m}C_{l]}^a$	Expression for the object of anholonomity $\Omega_{cb}^{\cdot\cdot a}$.

$\beta^{da} = \varepsilon^{dcb}\Omega_{cb}^{\cdot\cdot a}$	Expression for the density of newly created free surface.
(κa)	State associated with surface dislocation array.
(k^{I1})	Coherent dislocated interface of either the grain boundary or two-phase type.
(a^{I1})	Torn dislocated interface of either the grain boundary or two-phase type.
$Q_{ml}^{\cdot\cdot k} = -2g^{kn}\partial_m e_{ln}$	Nonvanishing of the $Q_{ml}^{\cdot\cdot k}$ tensor is synonymous within nonmetric geometry.
(κ^{I})	Coherent dislocation-free interface existing between two phases or two differently oriented bodies.
$a_{oc} = \sqrt{N^2+M^2}\ a_o$	Expression for coincidence site lattice unit cell length a_{oc} in terms of crystal lattice unit cell length a_o where N and M are integers.
$\tan\theta = \frac{N}{M}$	Relationship for angle of misfit of a grain boundary θ.
(k^{I3c})	Coherent dislocated interface of either the grain boundary or two-phase type represented in terms of a common coincidence site lattice.
$a_{ocs} = \frac{a_{oc}}{N^2+M^2}$	Expression for coincidence site lattice sublattice unit cell a_{ocs} expressed in terms of the coincidence site lattice unit cell a_{oc}.
$a_{oca} = \sqrt{o^2+p^2}\ a_{oc}$	Expression for the coincidence site lattice unit cell a_{oca} associated with an asymmetric boundary where o and p are integers.
$\tan\phi = \frac{o}{p}$	Rotation for angle of tilt ϕ of an asymmetric boundary.
$A_K^{a^6} = \delta_L^{a^6}\delta_K^{\kappa^6}B_{\kappa^6}^{L}$	Relation to show the equivalence between the $A_K^{a^6}$ and $B_{\kappa^6}^{L}$ distortions where $\delta_L^{a^6}$ and $\delta_K^{\kappa^6}$ are the Kronecker deltas.
$b^a = \int_s \Omega_{cb}^{\cdot\cdot a}\, dF^{cb}$	Surface integral that gives the closure failure associated with the creation of a free surface.
$b^{\kappa'} = \frac{1}{2}\int_s Q_{\mu'\lambda'}^{\cdot\cdot\kappa'}\, dF^{\mu'\lambda'}$	Surface integral that gives the virtual dislocation content associated with a given Burgers circuit.
u_n	Elastic displacent (linearized).
$e_{\kappa\lambda} = \frac{1}{2}\left(\frac{\partial u_\kappa}{\partial x_\lambda} + \frac{\partial u_\lambda}{\partial x_\kappa}\right)$	Expression for the elastic strain $e_{\kappa\lambda}$ in linearized form.
$\beta_{\kappa\lambda} = \frac{\partial u_\kappa}{\partial x_\lambda}$	Elastic distortion tensor $\beta_{\kappa\lambda}$ in linearized form.
$\beta_{\kappa\lambda} = e_{\kappa\lambda} + \omega_{\kappa\lambda}$	Expression of the linearized elastic distortion in terms of a symmetric $e_{\kappa\lambda}$ and an asymmetric $\omega_{\kappa\lambda}$ part.

$\delta_\lambda^K - \beta_\lambda^K = A_\lambda^K$	Relation between the distortions β_λ^K and A_λ^K.
$\beta_{ab}^P = \frac{\partial u_a}{\partial x_b}$	Plastic distortion tensor β_{ab}^P in linearized form.
$\beta_{ab}^P = e_{ab}^P + \omega_{ab}^P$	Expression of the linearized plastic distortion in terms of a symmetric e_{ab}^P and an asymmetric ω_{ab}^P part.
$\beta_{\kappa\lambda}^T = \beta_{\kappa\lambda} + \beta_{\kappa\lambda}^P$	Total linearized $\beta_{\kappa\lambda}^T$ expressed in terms of the elastic and plastic parts.
$4(\partial_{\mu\rho} e_{\kappa\lambda})_{[\mu\lambda][\rho\kappa]} \equiv \partial_{\mu\rho} e_{\kappa\lambda} - \partial_{\lambda\rho} e_{\kappa\mu} + \partial_{\lambda\kappa} e_{\rho\mu} - \partial_{\mu\kappa} e_{\rho\lambda} = 0$	Compatability conditions for elastic strain in the linear approximation.
$R_{\mu\lambda\rho\kappa} \equiv 2(\partial_{\mu\rho} g_{\kappa\lambda})_{[\mu\lambda][\rho\kappa]} = \frac{1}{2}(\partial_{\mu\rho} g_{\kappa\lambda} - \partial_{\lambda\rho} g_{\kappa\mu} + \partial_{\lambda\kappa} g_{\rho\mu} - \partial_{\mu\kappa} g_{\rho\lambda})$	Riemann–Christoffel curvature tensor of the first kind in the linear approximation.
$4(\partial_{\mu\rho} e_{\kappa\lambda})_{[\mu\lambda][\rho\mu]} = R_{\mu\lambda\rho\kappa}$	Alternate formulation of the Riemann–Christoffel curvature tensor of the first kind in the linear approximation.
$b_k = -\oint \beta_{kl}^P \, dx_l$	Line integral which gives the closure failure associated with a given Burgers circuit in the linearized approximation.
$S_{jlk} = \partial_{[j} \beta_{l]k}^P$	Expression for the torsion tensor S_{jlk} in the linear approximation.
$b_k = -\int_s S_{jlk} \, dF_{jl}$	Surface integral that gives the closure failure associated with the dislocation content of a given Burgers circuit in the linear approximation.
$\alpha_{ik} = -\varepsilon_{ijl} S_{jlk}$	Expression for the dislocation density tensor α_{ik} in the linear approximation.
$b_k = \int_s \alpha_{ik} \, dF_i$	Alternate expression for the closure failure in terms of the dislocation density tensor in the linearized form.
$dC_k = -\Gamma_{mlk} C_l \, dx_m$	Linearized form of the equation for parallel displacement.
$x\delta(x)$	Relation satisfied by the Dirac delta function.
$-G_{mlrk} = \partial_{mr} e_{kl} - \partial_{lr} e_{km} + \partial_{lk} e_{rm} - \partial_{mk} e_{rl}$	Incompatability tensor G_{mlrk} associated with a dislocated body in the linear approximation.
$\sigma_{kl} = C_{klij} e_{ij}$	Hooke's law relating stress σ_{kl} and strain e_{ij} in the linear approximation.
$\eta_{pq} = \frac{1}{4} \varepsilon_{pml} \varepsilon_{grk} G_{mlrk}$	Alternate form of the incompatability tensor η_{pq}.
$\sigma_{kl} = -\varepsilon_{kij} \varepsilon_{lmn} \partial_{im} \psi_{jn}$	Representation of stress in terms of the stress function ψ_{jn}.
$\chi_{ij} = -\frac{1}{8\Pi} \int \eta_{ij}(\boldsymbol{r}') \lvert \boldsymbol{r} - \boldsymbol{r}' \rvert \, dV'$	Solution for the stress function in terms of the Green's function $-1/8\Pi\lvert \boldsymbol{r} - \boldsymbol{r}' \rvert$.

$\int_{-\infty}^{y-\varepsilon} + \int_{y+\varepsilon}^{\infty} \frac{\mu b_s}{2\Pi} \frac{f(y')\,dy'}{(y-y')}$	Value of the stress σ_{xz} at some point y due to all the other dislocations in a continuous distribution of screw type dislocation where $b_s f(y')\,dy'$ is the Burgers vector of the dislocation lying between y' and $y'+dy'$.
$f(y) = \frac{b_l}{\Pi b_s} \frac{d}{(d^2+y^2)}$	Distribution function associated with a continuous array of surface dislocations associated with a screw dislocation of strength b_l located at a distance of d from the free surface of a semi-infinite body.
$\int_{-\infty}^{+\infty} b_s f(y)\,dy = b_l$	Statement of the conservation law associated with a crystal lattice dislocation and its surface dislocation array.
(k^D)	States comprised of a single disclination.
$R_{nml}^{\cdot\cdot\cdot k} = \partial_n \Gamma_{ml}^k - \partial_m \Gamma_{nl}^k + \Gamma_{nr}^k \Gamma_{ml}^r - \Gamma_{mr}^k \Gamma_{nl}^r$	Expression for the Riemann–Christoffel curvature tensor of the second kind $R_{nml}^{\cdot\cdot\cdot k}$.
$b^k = \frac{1}{2} \int_s R_{nml}^{\cdot\cdot\cdot k} C^l\,dF^{nm}$	Closure failure associated with the parallel transport of a vector C^l about a given Burgers circuit.
$\kappa_{mkl}^{p} = \partial_m \beta_{kl}^{p\omega}$	Definition of the bend-twist tensor κ_{mkl}^{p} in the linear approximation.
$E_l = E_s + E_I$	Elastic strain energy released by the formation of an elastic crack where E_s and E_I are the self and interaction energies of the crack dislocations.
$E_l^p = E_s + E_I - E_f$	Elastic strain energy released by the formation of a plastic crack where E_f is the frictional energy associated with the movement of crystal lattice dislocations in the plastic zone of the crack.
$R_{0G} = \frac{4\gamma\mu}{\Pi(1-\nu)\sigma_{xy}^2}$	Griffith's condition for the formation of an elastic crack of size R_{0G}.
$E_s = \frac{\mu b^2}{4\Pi(1-\nu)} \ln\left(\frac{R}{r_o}\right)$	Self energy per unit length of an infinite straight edge-type dislocation.
$E_I = -\frac{\mu b^2}{2\Pi(1-\nu)}\left[\ln\frac{R}{d} - \sin^2\theta\right]$	Interaction energy per unit length between two parallel infinite straight edge-type dislocations separated by a distance d.

Unified Theory of the Mechanical Behavior of Matter

CHAPTER 1

Introduction

Ever since my initiation into the fields of elasticity, plasticity, and dislocation theory during graduate training in the early fifties, I have felt a certain perplexity regarding these disciplines. On the one hand, those who deal with classical elasticity pay virtually no attention to dislocation theory (Sneddon and Berry, 1958; Truesdell and Toupin, 1960). These individuals fall mainly in the disciplines of applied mathematics, civil and mechanical engineering, as well as theoretical and applied mechanics. Similarly, those who concern themselves with classical plasticity also pay virtually no attention to dislocation theory (Freudenthal, 1958; Hill, 1950; Prager, 1959; Hoffman and Sachs, 1953). In fact, essentially all of these theories have as their starting point the classical theory of elasticity. It is therefore not surprising that most of those working in this field are to be found within the same general disciplines as those involved with classical elasticity.

We now come to the field of dislocation theory. In spite of the fact that the dislocation was originally formulated by Volterra (1907) as an elastic entity, in their eagerness to develop a stronger theory of plasticity, a number of early investigators (Orowan, 1934; Polanyi, 1934; Taylor, 1934) made the association between dislocations and plasticity almost synonymous. This has persisted up to the present day with unfortunate consequences, as the following study shows. Most of the individuals dealing with the theory of dislocations are to be found in the disciplines of applied mathematics, theoretical and applied physics, metallurgical, civil, and mechanical engineering, as well as materials science. In spite of the obvious inherently close relationship between the fields of elasticity, plasticity, and dislocation theory, except for the first two disciplines, there is little if any interaction between the various groups. In fact, we may go so far as to say that there is even open hostility between those dealing with classical elasticity and plasticity on the one hand, and those working with dislocation theory. The question that now naturally arises is how could such a perplexing state of affairs have arisen in such an important area of science and engineering? As is shown in the following study, those who study classical elasticity are to be faulted for their inwardness, which is reflected in their refusal to employ dislocation theory in spite of the fact that it was born within their discipline. The problems with classical plasticity result from a too strong dependence upon classical elasticity, with the result that even a basic definition of pure plastic distortion is overlooked. Finally, in spite of its inherent beauty and power, dislocation theory has suffered and is still suffering from an excess of imprecise, and even incorrect or incomplete, definitions, concepts, and theories. It is the confusion resulting from this impreciseness that has kept many of the leading thinkers in other disciplines from entering into the field of dislocation theory. In fact, even workers in metallurgy and materials science, who were among the last holdouts and later among the most enthusiastic champions of dislocation theory, have now given it low priority because it does not measure up to their initial expectations. This is reflected in recent attitudes taken by certain research funding agencies (Marcinkowski, 1977a).

In my frustration with this state of affairs and with a determination to rectify the situation, I decided to cast aside the conventional thinking and, from a few simple concepts, begin to rethink the mechanical behavior of matter.

CHAPTER 2

Elastic Distortions

The meaning of a pure elastic distortion is relatively well-defined so that this chapter is in a sense a general review of the topic. No restrictions are made on the material with respect to the particular constitutive relations between stress and strain so that the analysis applies to such materials as metals, which of necessity involve small strains, and to rubberlike substances, where the strains are correspondingly large. This makes it necessary to use the more generalized covariant and contravariant index notation for the various tensor quantities. Also for simplicity, the various strains to be considered are taken to be uniform.

Before formulating any definition of strain, it is first necessary to define an element of distance or length ds. This can be done most conveniently by considering the unstrained reference state crystal shown in Figure 2.1*a*. Capital Latin letters, that is K, L, M, and so on, are used to denote this reference crystal. An element of length can now be defined as (Fung, 1965; Sokolnikoff, 1964; Flügge, 1972)

$$\left(\underset{K}{ds}\right)^2 = a_{KL}\, dx^K dx^L \tag{2.1}$$

where a_{KL} is the metric tensor associated with the (K) state, while dx^K are the components of length measured with respect to the (K) state. It is important to note that we have been very careful to distinguish between covariant components, that is, a_{KL}, written as subscripts, and contravariant components, dx^K, written as superscripts. The metric tensor a_{KL} can easily be found from the following relationship:

$$a_{KL} = \mathbf{e}_K \cdot \mathbf{e}_L \tag{2.2}$$

where $\mathbf{e}_K$ are the unit base vectors associated with the (K) state. Since the

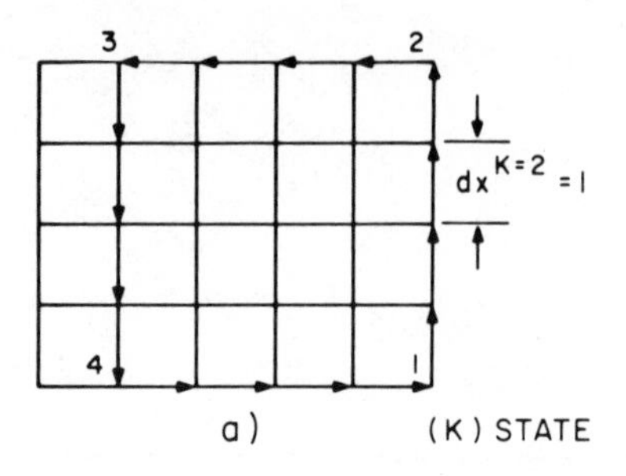

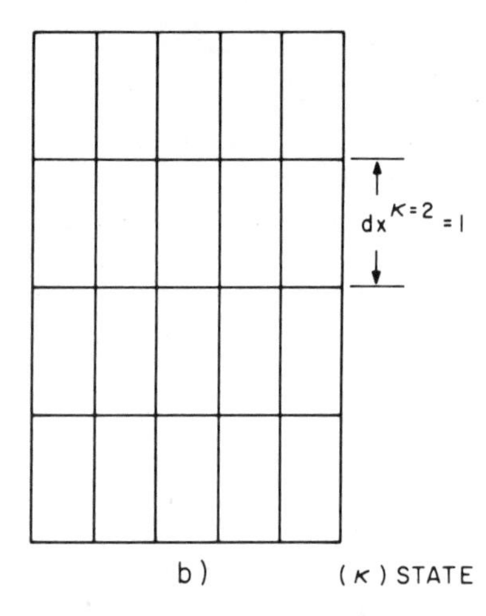

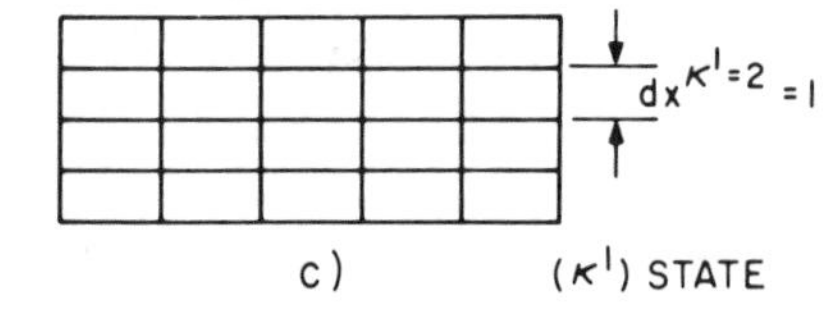

Figure 2.1 *a*) Unstrained reference state. Elastically strained states of uniaxial *b*) tension, *c*) compression.

(K) state is Cartesian, it follows that

$$a_{KL}=\delta_{KL} \tag{2.3}$$

where δ_{KL} is the Kronecker delta defined such that

$$\delta_{KL}=\begin{cases}1 & \text{if } K=L\\ 0 & \text{if } K\neq L\end{cases} \tag{2.4}$$

The (K) state of Figure 2.1a can now be given an uniaxial elongation to produce the elastically strained state shown in Figure 2.1b. Such elastically strained states are denoted by lowercase Greek letters, that is, κ, λ, μ, and so on. An element of length in the (κ) state must now be written as follows:

$$\left(\underset{\kappa}{ds}\right)^2 = g_{\kappa\lambda}\, dx^{\kappa}\, dx^{\lambda} \tag{2.5}$$

where $g_{\kappa\lambda}$ is the metric tensor associated with the (κ) state. It may be written as

$$g_{\kappa\lambda} = \mathbf{e}_{\kappa} \cdot \mathbf{e}_{\lambda} \tag{2.6}$$

where $\mathbf{e}_{\kappa}$ are the base vectors associated with the (κ) state. It is important to note that they are no longer unit vectors. This can be seen by observing that

$$\mathbf{e}_{\kappa} = B_{\kappa}^{K} \mathbf{e}_{K} \tag{2.7a}$$

or

$$\mathbf{e}_{K} = B_{K}^{\kappa} \mathbf{e}_{\kappa} \tag{2.7b}$$

where the B_{κ}^{K} may be termed elastic distortion tensors whose components can be written out in full as

$$B_{\kappa}^{K} = \begin{pmatrix} B_1^1 & B_1^2 & B_1^3 \\ B_2^1 & B_2^2 & B_2^3 \\ B_3^1 & B_3^2 & B_3^3 \end{pmatrix} \tag{2.8}$$

The elastic distortion tensors obey the following relationship:

$$B_{\kappa}^{K} B_{K}^{\lambda} = \delta_{\kappa}^{\lambda} \tag{2.9}$$

In the case of the (κ) state shown in Figure 2.1*b* it follows that

$$B_1^1 = B_3^3 = 1 \tag{2.10a}$$

while

$$B_2^2 = 2 \tag{2.10b}$$

with all other components equal to zero. Combining Eqs. 2.6 and 2.7, we can now write

$$g_{\kappa\lambda} = \mathbf{e}_{\kappa} \cdot \mathbf{e}_{\lambda} = B_{\kappa}^{K} B_{\lambda}^{L} \mathbf{e}_{K} \cdot \mathbf{e}_{L} = B_{\kappa}^{K} B_{\lambda}^{L} \delta_{KL} = B_{\kappa}^{K} B_{\lambda}^{K} \tag{2.11}$$

In view of Eq. 2.10, the above relation yields for Figure 2.1*b*

$$g_{11}=g_{33}=1 \tag{2.12a}$$

and

$$g_{22}=4 \tag{2.12b}$$

while all other components vanish.

It is also important to note that in our definitions of distance, embodied in Eqs. 2.1 and 2.5, is the assumption that the coordinates maintain their same values in the (K) and (κ) states, that is, they are said to be dragged along (Schouten, 1954). This means that if the coordinates in these two states are connected by the distortion tensor A_K^κ then

$$dx^\kappa = A_K^\kappa dx^K \tag{2.13a}$$

or

$$dx^K = A_\kappa^K dx^\kappa \tag{2.13b}$$

where

$$A_K^\kappa = \delta_K^\kappa \tag{2.14a}$$

and

$$A_\kappa^K = \delta_\kappa^K \tag{2.14b}$$

This is illustrated schematically in Figure 2.1*a* and *b* for the components $dx^{K=2}$ and $dx^{\kappa=2}$ where

$$dx^{K=2} = dx^{\kappa=2} = 1 \tag{2.15}$$

We are now in a position to formulate an exact definition of elastic strain. In particular, let us first write with the aid of Eqs. 2.1, 2.5, and 2.13

$$\left(\underset{\kappa}{ds}\right)^2 - \left(\underset{K}{ds}\right)^2 = \left(g_{\kappa\lambda} - a_{KL}A_\kappa^K A_\lambda^L\right) dx^\kappa dx^\lambda \tag{2.16}$$

The strain tensor is now defined as

$$e_{\kappa\lambda} = \tfrac{1}{2}\left(g_{\kappa\lambda} - a_{KL}A_\kappa^K A_\lambda^L\right) \tag{2.17}$$

which, in view of the dragging conditions imposed by Eq. 2.14, becomes

$$e_{\kappa\lambda} = \tfrac{1}{2}(g_{\kappa\lambda} - a_{\kappa\lambda}) \tag{2.18}$$

Thus, the elastic strain is given simply in terms of the metric tensors of the initial and final states. In view of Eqs. 2.3 and 2.12, for the $(K)\rightarrow(\kappa)$ distortion, Eq. 2.18 yields

$$e_{22} = \tfrac{3}{2} \tag{2.19}$$

while all other components of $e_{\kappa\lambda}$ vanish.

In order to obtain a more general insight into the elastic distortion of solids, let us consider several additional types of deformation. In particular, consider the (κ^1) state shown in Figure 2.1*c* which is generated from the (K) state by a uniaxial compression. The nonvanishing components of the elastic distortion tensor $B_{\kappa^1}^{K}$ for this case are

$$B_1^1 = B_3^3 = 1 \tag{2.20a}$$

and

$$B_2^2 = \tfrac{1}{2} \tag{2.20b}$$

Again, as in all subsequent elastic distortions, the coordinates are dragged so that

$$dx^{\kappa^1} = \delta_K^{\kappa^1}\, dx^K \tag{2.21}$$

The nonvanishing components of the metric tensor $g_{\kappa^1\lambda^1}$ can be found from an expression of the type given by Eq. 2.11 to be

$$g_{11} = g_{33} = 1 \tag{2.22a}$$

while

$$g_{22} = \tfrac{1}{4} \tag{2.22b}$$

These in turn allow the nonvanishing component of the strain tensor to be written as

$$e_{22} = -\tfrac{3}{8} \tag{2.23}$$

where an equation of the type given by Eq. 2.18 has been employed.

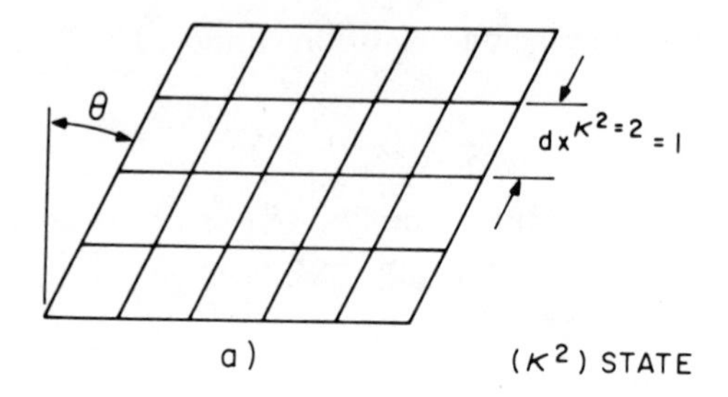

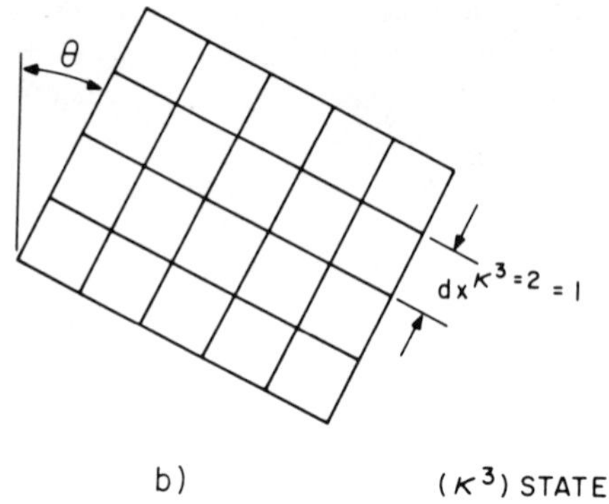

Figure 2.2 Elastically strained states of *a*) simple shear, *b*) rigid rotation.

The state of simple shear (κ^2) shown in Figure 2.2*a* may next be generated from the (K) state via the elastic distortion $B^K_{\kappa^2}$ where the only nonvanishing components are

$$B^1_1 = B^2_2 = B^3_3 = 1 \tag{2.24a}$$

and

$$B^1_2 = \tan\theta = \tfrac{1}{2} \tag{2.24b}$$

where the angle θ is as indicated in Figure 2.2*a*. The above distortion allows the metric tensor for the (κ^2) state to found as

$$g_{\kappa^2\lambda^2} = B^K_{\kappa^2} B^K_{\lambda^2} = \begin{bmatrix} 1 & \tan\theta & 0 \\ \tan\theta & (\tan^2\theta + 1) & 0 \\ 0 & 0 & 1 \end{bmatrix} \tag{2.25}$$

which can be used to obtain the following strain tensor:

$$e_{\kappa^2\lambda^2} = \begin{bmatrix} 0 & \frac{1}{2}\tan\theta & 0 \\ \frac{1}{2}\tan\theta & \frac{1}{2}\tan^2\theta & 0 \\ 0 & 0 & 0 \end{bmatrix} \tag{2.26}$$

The (K) state of Figure 2.1a can next be rigidly rotated about the z axis by θ to generate the (κ^3) state shown in Figure 2.2b. In this case the distortion tensor becomes

$$B_{\kappa^3}^{K} = \begin{bmatrix} \cos\theta & -\sin\theta & 0 \\ \sin\theta & \cos\theta & 0 \\ 0 & 0 & 1 \end{bmatrix} \tag{2.27}$$

which yields the following metric tensor:

$$g_{\kappa^3\lambda^3} = B_{\kappa^3}^{K} B_{\lambda^3}^{K} = \delta_{\kappa^3\lambda^3} = a_{\kappa^3\lambda^3} \tag{2.28}$$

where the last equality follows from Eq. 2.3. It is obvious from Eq. 2.28 that

$$e_{\kappa^3\lambda^3} = \tfrac{1}{2}(g_{\kappa^3\lambda^3} - a_{\kappa^3\lambda^3}) = 0 \tag{2.29}$$

which is a well-known result for a rigid rotation.

Still more complex elastic distortions can be considered. For example, consider the (κ^4) state of Figure 2.3a which is generated from the (K) state by the following elastic distortion tensor:

$$B_{\kappa^4}^{K} = \begin{bmatrix} 1 & -\tan\theta & 0 \\ \tan\theta & 1 & 0 \\ 0 & 0 & 1 \end{bmatrix} \tag{2.30}$$

The corresponding metric tensor becomes

$$g_{\kappa^4\lambda^4} = B_{\kappa^4}^{K} B_{\lambda^4}^{K} = \begin{bmatrix} (1+\tan^2\theta) & 0 & 0 \\ 0 & (1+\tan^2\theta) & 0 \\ 0 & 0 & 1 \end{bmatrix} \tag{2.31}$$

from which the corresponding strain tensor is found to be

$$e_{\kappa^4\lambda^4} = \begin{bmatrix} \frac{1}{2}\tan^2\theta & 0 & 0 \\ 0 & \frac{1}{2}\tan^2\theta & 0 \\ 0 & 0 & 0 \end{bmatrix} \tag{2.32}$$

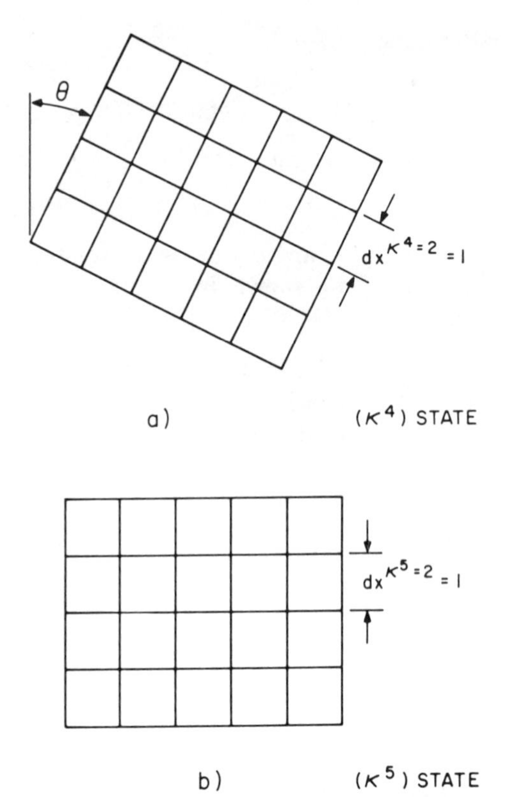

Figure 2.3 Elastically strained states of *a*) double shear, *b*) biaxial tension.

It is apparent from the vanishing of the off-diagonal elements in the last two equations that the $(K)\rightarrow(\kappa^4)$ distortion is simply a volume distortion. This can be seen by rewriting Eq. 2.30 as

$$B_{\kappa^4}^{K} = B_{\kappa^4}^{\kappa^5} B_{\kappa^5}^{K} \tag{2.33}$$

where

$$B_{\kappa^4}^{\kappa^5} = \delta_{K}^{\kappa^5} \delta_{\kappa^4}^{\kappa^3} B_{\kappa^3}^{K} \tag{2.34a}$$

as given by Eq. 2.27, while

$$B_{\kappa^5}^{K} = \begin{pmatrix} \dfrac{1}{\cos\theta} & 0 & 0 \\ 0 & \dfrac{1}{\cos\theta} & 0 \\ 0 & 0 & 1 \end{pmatrix} \tag{2.34b}$$

where state (κ^5) is the biaxially strained state shown in Figure 2.3*b*. Thus, the (κ^4) state can be visualized as being generated from the (K) state by a biaxial expansion followed by a rigid rotation. The metric tensor associated with the (κ^5) state is simply

$$g_{\kappa^5\lambda^5} = B_{\kappa^5}^{K} B_{\lambda^5}^{K} = \begin{pmatrix} \dfrac{1}{\cos^2\theta} & 0 & 0 \\ 0 & \dfrac{1}{\cos^2\theta} & 0 \\ 0 & 0 & 1 \end{pmatrix} \tag{2.35}$$

Comparison of the last expression with Eq. 2.31 shows that

$$g_{\kappa^5\lambda^5} = \delta_{\kappa^5}^{\kappa^4} \delta_{\lambda^5}^{\lambda^4} g_{\kappa^4\lambda^4} \tag{2.36}$$

It also follows from Eq. 2.32 that

$$e_{\kappa^5\lambda^5} = \delta_{\kappa^5}^{\kappa^4} \delta_{\lambda^5}^{\lambda^4} e_{\kappa^4\lambda^4} \tag{2.37}$$

The last two equations follow from the fact that a rigid rotation leaves both the metric and strain tensors unaltered.

A double shear can also be applied to the (K) state in which one of the shears is of opposite sense to the other so as to generate the (κ^6) state shown in Figure 2.4. This may be accomplished by means of the following elastic distortion tensor:

$$B_{\kappa^6}^{K} = \begin{pmatrix} 1 & \tan\theta & 0 \\ \tan\theta & 1 & 0 \\ 0 & 0 & 1 \end{pmatrix} \tag{2.38}$$

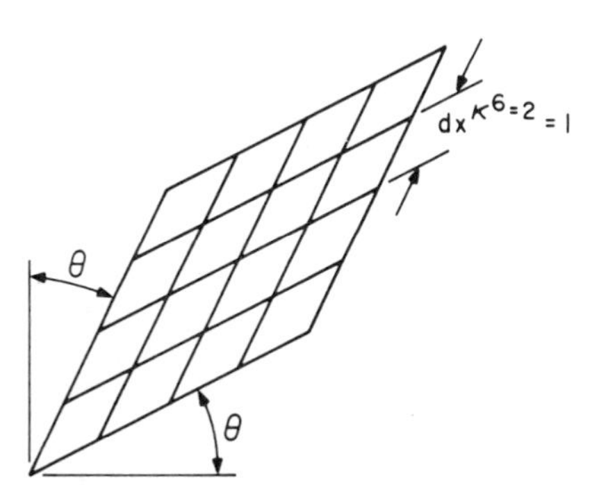

Figure 2.4 Elastically strained state of double shear, but in opposite sense.

For this particular state, the metric tensor becomes

$$g_{\kappa^6\lambda^6} = B_{\kappa^6}^{K} B_{\lambda^6}^{K} = \begin{bmatrix} (1+\tan^2\theta) & 2\tan\theta & 0 \\ 2\tan\theta & (1+\tan^2\theta) & 0 \\ 0 & 0 & 1 \end{bmatrix} \tag{2.39}$$

which yields a strain tensor given by

$$e_{\kappa^6\lambda^6} = \begin{bmatrix} \frac{1}{2}\tan^2\theta & \tan\theta & 0 \\ \tan\theta & \frac{1}{2}\tan^2\theta & 0 \\ 0 & 0 & 0 \end{bmatrix} \tag{2.40}$$

Note that, unlike the distortion tensor given by Eq. 2.30 which generated the (κ^4) state, it is not possible to decompose the distortion tensor $B_{\kappa^6}^{K}$ given by Eq. 2.38 into a rigid rotation plus a pure volume distortion.

The volume V associated with any of the states considered thus far may be written as (Sokolnikoff, 1964; Flügge, 1972)

$$V_{(\kappa)} = \mathbf{e}_1 \times \mathbf{e}_2 \cdot \mathbf{e}_3 = \det\left(B_{\kappa}^{K}\right) \tag{2.41}$$

where det signifies determinant. The volumes of the various states are thus found to be

$$V_{(K)} = V_{(\kappa^2)} = V_{(\kappa^3)} = 1 \tag{2.42a}$$

$$V_{(\kappa)} = 2 \tag{2.42b}$$

$$V_{(\kappa^1)} = \tfrac{1}{2} \tag{2.42c}$$

$$V_{(\kappa^4)} = V_{(\kappa^5)} = \frac{1}{\cos^2\theta} \tag{2.42d}$$

$$V_{(\kappa^6)} = 1 - \tan^2\theta \tag{2.42e}$$

Thus, the distortions from state (K) to states (κ^2) and (κ^3) lead to no volume change, whereas those from state (K) to states (κ^1) and (κ^6) lead to a volume decrease, while those from state (K) to states (κ), (κ^4), and (κ^5) give rise to a volume increase.

It is apparent from Eqs. 2.25 and 2.39 that the metric tensors associated with states (κ^2) and (κ^6), respectively, are nondiagonal. They can be diagonalized by first writing the conditions for principal directions, that is (Wayman, 1964; Coburn, 1970; Schouten, 1951),

$$g_{\kappa^2\lambda^2} x^{\lambda^2} = \lambda a_{\kappa^2\lambda^2} x^{\lambda^2} \tag{2.43a}$$

In view of Eq. 2.3 this becomes

$$g_{\kappa^2\lambda^2}x^{\lambda^2}=\lambda\delta_{\kappa^2\lambda^2}x^{\lambda^2} \tag{2.43b}$$

where λ are the eigenvalues. The above equation has a solution when

$$\det(g_{\kappa^2\lambda^2}-\lambda\delta_{\kappa^2\lambda^2})=0 \tag{2.44}$$

Expansion of the above equation yields

$$\begin{aligned}\lambda^3=(g_{11}+g_{22}+g_{33})\lambda^2+(g_{11}g_{22}+g_{22}g_{33}+g_{33}g_{11}-g_{12}g_{21}\\ -g_{23}g_{32}-g_{31}g_{13})\lambda-\det g_{\kappa^2\lambda^2}=0\end{aligned} \tag{2.45}$$

The three roots of the above equation are the eigenvalues of $g_{\kappa^2\lambda^2}$ which may be written as

$$g_{\kappa^D\lambda^D}=\begin{bmatrix} g_{11} & 0 & 0 \\ 0 & g_{22} & 0 \\ 0 & 0 & g_{33}\end{bmatrix}=\begin{bmatrix} \underset{1}{\lambda} & 0 & 0 \\ 0 & \underset{2}{\lambda} & 0 \\ 0 & 0 & \underset{3}{\lambda}\end{bmatrix} \tag{2.46}$$

so that the metric tensor is diagonalized. The diagonalized state may be referred to as (κ^D). Since the terms in parentheses in Eq. 2.45 are invariants, it follows that

$$\underset{1}{\lambda}+\underset{2}{\lambda}+\underset{3}{\lambda}=g_{11}+g_{22}+g_{33} \tag{2.47a}$$

$$\begin{aligned}\underset{1}{\lambda}\underset{2}{\lambda}+\underset{2}{\lambda}\underset{3}{\lambda}+\underset{3}{\lambda}\underset{1}{\lambda}=g_{11}g_{22}+g_{22}g_{33}+g_{33}g_{11}\\ -g_{12}g_{21}-g_{23}g_{32}-g_{31}g_{13}\end{aligned} \tag{2.47b}$$

$$\underset{1}{\lambda}\underset{2}{\lambda}\underset{3}{\lambda}=\det g_{\kappa^2\lambda^2} \tag{2.47c}$$

In view of Eq. 2.25, the above relations yield

$$\underset{1}{\lambda}+\underset{2}{\lambda}+\underset{3}{\lambda}=\tan^2\theta+3 \tag{2.48a}$$

$$\underset{1}{\lambda}\underset{2}{\lambda}+\underset{2}{\lambda}\underset{3}{\lambda}+\underset{3}{\lambda}\underset{1}{\lambda}=\tan^2\theta+3 \tag{2.48b}$$

$$\underset{1}{\lambda}\underset{2}{\lambda}\underset{3}{\lambda}=\tan^2\theta+1 \tag{2.48c}$$

Solution of these equations gives

$$\underset{1}{\lambda} = \frac{2}{(\tan^2\theta + 2) \pm \sqrt{(\tan^2\theta + 2)^2 - 4}} = 0.577 \tag{2.49a}$$

$$\underset{2}{\lambda} = \frac{(\tan^2\theta + 2) \pm \sqrt{(\tan^2\theta + 2) - 4}}{2} = 1.74 \tag{2.49b}$$

$$\underset{3}{\lambda} = 1 \tag{2.49c}$$

where again, $\tan\theta$ was chosen as $1/2$. These results are most readily illustrated by reference to Figure 2.5 which shows the distortion of a unit circle into an ellipsoid by the distortion tensor B_κ^K. This is merely another representation of Figure 2.2*a*. The roots $\underset{1}{\lambda}$ and $\underset{2}{\lambda}$ are simply the lengths

$$(ds)^2 = \lambda_{11} x^{\kappa^D = 1} x^{\kappa^D = 1} \tag{2.50a}$$

$$(ds)^2 = \lambda_{22} x^{\kappa^D = 2} x^{\kappa^D = 2} \tag{2.50b}$$

in Figure 2.5. Once again, the coordinates have been dragged. It is thus obvious from Eqs. 2.46 and 2.49 that a simple shear is equivalent to a simultaneous uniaxial tension and a compression at right angles to one another. The directions or eigenvectors corresponding to $\underset{1}{\lambda}$, $\underset{2}{\lambda}$, and $\underset{3}{\lambda}$ are easily found by solving Eq. 2.43b which in expanded form yields

$$\begin{aligned} g_{11}x^1 + g_{12}x^2 + g_{13}x^3 &= \lambda\delta_{11}x^1 \\ g_{21}x^1 + g_{22}x^2 + g_{23}x^3 &= \lambda\delta_{22}x^2 \\ g_{31}x^1 + g_{32}x^2 + g_{33}x^3 &= \lambda\delta_{33}x^3 \end{aligned} \tag{2.51}$$

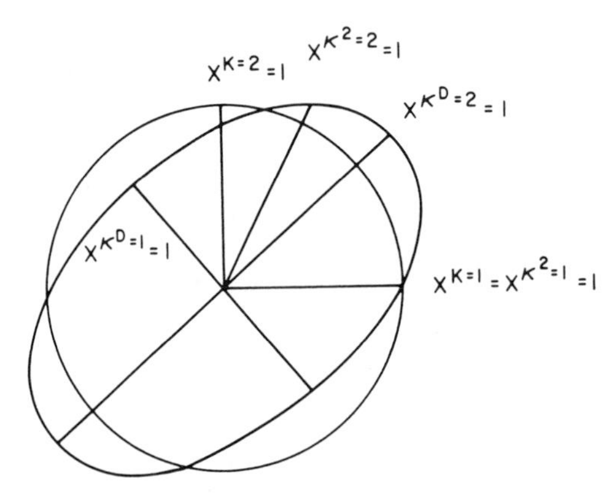

Figure 2.5 Elastic distortion of a circle by simple shear.

Each of the eigenvalues given by Eq. 2.46, when substituted into the above equations, yield the corresponding components of the eigenvector x^{κ^D}. For example, for $\underset{1}{\lambda}$, we obviously have $\underset{1}{x}^3 = 0$. Choosing $\underset{1}{x}^1$ equal to some constant k, the first of Eq. 2.51 yields

$$\underset{1}{x}^2 = \frac{\left(1 - \underset{1}{\lambda}\right)k}{\tan\theta} \tag{2.52}$$

We thus have for the components of $\underset{1}{x}$, $\underset{2}{x}$, and $\underset{3}{x}$

$$\underset{1}{x} = \left(k, -\frac{\left(1 - \underset{1}{\lambda}\right)k}{\tan\theta}, 0\right) \tag{2.53a}$$

$$\underset{2}{x} = \left[\frac{k\tan\theta}{\left(1 - \underset{2}{\lambda}\right)}, k, 0\right] \tag{2.53b}$$

$$\underset{3}{x} = (1, 0, 0) \tag{2.53c}$$

Each of these vectors can in turn be written in normalized form to yield $x^{\kappa^D=1}$, $x^{\kappa^D=2}$, and $x^{\kappa^D=3}$, respectively. We could also write

$$x^{\kappa^D} = C_K^{\kappa^D} x^K \tag{2.54}$$

where again x^{κ^D} and x^K are of unit length while $C_K^{\kappa^D}$ is a coordinate transformation whose components are given by the normalized forms of Eq. 2.53. It obviously corresponds to a rigid rotation of axes.

It is now possible to define a distortion tensor $B_{\kappa^D}^K$ which generates the (κ^D) state directly from the reference state. In particular,

$$B_{\kappa^D}^K = C_{\kappa^7}^K B_{\kappa^D}^{\kappa^7} \tag{2.55}$$

where $C_{\kappa^7}^K$ is simply the coordinate transformation given by

$$C_{\kappa^7}^K = \delta_{\kappa^7}^L \delta_{\kappa^D}^K C_L^{\kappa^D} \tag{2.56}$$

as given by Eq. 2.54. The distortion tensor $B_{\kappa^D}^{\kappa^7}$ is simply given by

$$B_{\kappa^D}^{\kappa^7} = \begin{bmatrix} \underset{1}{\lambda} & 0 & 0 \\ 0 & \underset{2}{\lambda} & 0 \\ 0 & 0 & 1 \end{bmatrix} \tag{2.57}$$

The above distortion can in turn be used to obtain $g_{\kappa^D\lambda^D}$ and $e_{\kappa^D\lambda^D}$ associated with the (κ^D) state. It is clear that these strain tensors are related to those in the (κ^2) state simply by a coordinate transformation. In particular

$$g_{\kappa^D\lambda^D} = C_{\kappa^D}^{\kappa^2} C_{\lambda^D}^{\lambda^2} g_{\kappa^2\lambda^2} \tag{2.58a}$$

while

$$e_{\kappa^D\lambda^D} = C_{\kappa^D}^{\kappa^2} C_{\lambda^D}^{\lambda^2} e_{\kappa^2\lambda^2} \tag{2.58b}$$

REVIEW

Six basic types of elastic distortion have been considered, that is, uniaxial tension and compression, simple shear, double shear of the same, and opposite sense and biaxial tension. More complex strain configurations can be formed from combinations of these basic distortions. These elastically strained states (κ) may be generated from an unstrained reference crystal (K) by means of a suitable distortion tensor B_κ^K. This distortion tensor can in turn be used to define a metric tensor $g_{\kappa\lambda}$ that can then be used to find the elastic strain tensor $e_{\kappa\lambda}$. The distortion tensor also connects the base vectors $\mathbf{e}_\kappa$ and $\mathbf{e}_K$ between the strained and unstrained states, respectively, and it is these base vectors that change in response to an elastic distortion, whereas the coordinates dx^κ remain unaltered, that is, they are said to be dragged. This is the very essence of an elastic distortion. As is shown in the following chapter, it is this property that uniquely distinguishes an elastic distortion from a plastic distortion. Finally, as Chapter 5 shows, the various states of elastic distortion considered in the present chapter can also be represented in terms of some suitable arrangement of surface dislocations. This is a very powerful concept in that it allows pure elastic distortions to be accurately represented in terms of various types of continuous dislocation distributions.

CHAPTER 3

Plastic Distortions

Although the concept of an elastic distortion, as discussed in the last chapter, has been known for a long time, the definition of plastic distortion has been somewhat more elusive. This is compounded by the fact that theories of plasticity, that is, the St. Venant, Tresca, von Mises, and so on, criteria are deeply rooted in elasticity theory. There seems to be no definition of a pure plastic distortion in which elastic distortions are absent. As is shown in subsequent chapters, such a definition is absolutely essential if one is to understand the deformation behavior of matter, in general. It is therefore the purpose of this chapter to formulate, for the first time, an all-inclusive concept of plastic deformation.

It is now possible to formulate a principle that states that for every elastic distortion there exists a corresponding pure plastic distortion (Marcinkowski, 1978a) These states of pure plastic distortion are designated by lowercase Latin letters at the front of the alphabet, that is a, b, c, and so on. For example, the plastic state corresponding to the uniaxial elastic distortion shown in Figure 2.1*b* is illustrated in Figure 3.1*a*. This state is denoted as (a) and may be generated from the (K) state by means of the plastic distortion tensor A_K^a which connects the coordinates in both states as follows:

$$dx^a = A_K^a dx^K \tag{3.1a}$$

or

$$dx^K = A_a^K dx^a \tag{3.1b}$$

where

$$A_a^K A_K^b = \delta_a^b \tag{3.2}$$

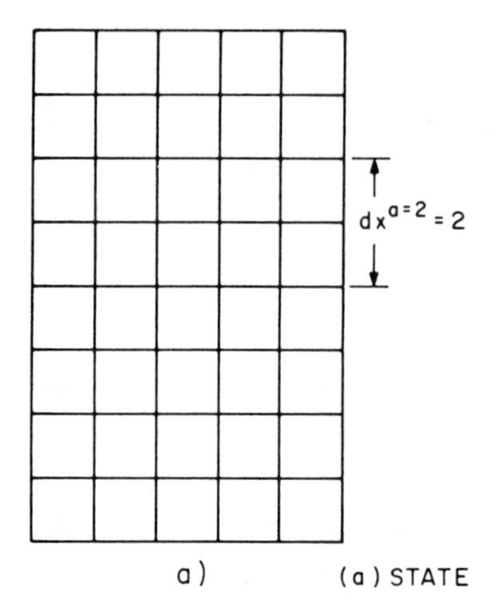

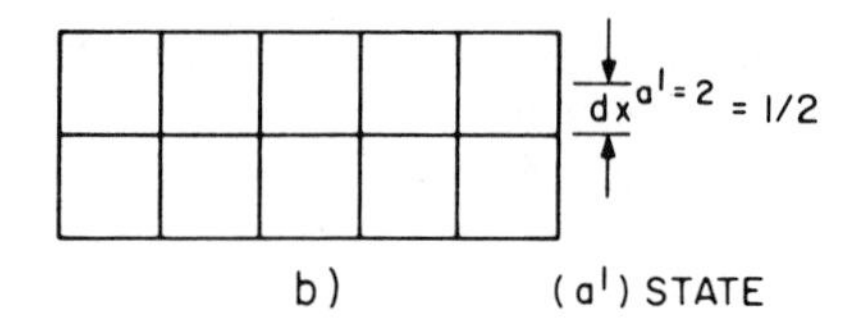

Figure 3.1 Plastically strained states of uniaxial *a*) tension, *b*) compression.

The nonvanishing components of A_K^a are

$$A_1^1 = A_2^2 = 1 \tag{3.3a}$$

while

$$A_2^2 = 2 \tag{3.3b}$$

It is thus apparent that, unlike the case for the elastic distortions, the coordinates are not dragged, that is

$$A_K^a \neq \delta_K^a \tag{3.4}$$

This is seen in Figure 3.1*a* where the coordinate $dx^{a=2}=2$ is generated from $dx^{K=2}=1$ in Figure 2.1*a*.

Since there is no elastic distortion associated with the $(K) \rightarrow (a)$ distortion, it follows that

$$B_K^a = \delta_K^a \tag{3.5a}$$

while

$$B_a^K = \delta_a^K \tag{3.5b}$$

This means, in view of the following relations:

$$\mathbf{e}_a = B_a^K \mathbf{e}_K \tag{3.6a}$$

and

$$\mathbf{e}_K = B_K^a \mathbf{e}_a \tag{3.6b}$$

that the base vectors are unchanged by the plastic distortion. Also, in view of

$$a_{ab} = \mathbf{e}_a \cdot \mathbf{e}_b = B_a^K B_b^K = \delta_{ab} \tag{3.7}$$

it follows that the metric tensors in the (K) and (a) states are identical, that is,

$$a_{ab} = a_{KL} \delta_a^K \delta_b^L \tag{3.8}$$

Thus we can now formulate the following important rule that states that whereas a pure elastic distortion is characterized by a change in the metric tensors but a dragging of coordinates, a pure plastic distortion is characterized by a change in coordinates but a dragging of the metric tensor (Marcinkowski, 1978b). It also follows that

$$A_a^K = B_L^\kappa \delta_\kappa^K \delta_a^L \tag{3.9a}$$

while

$$A_a^K B_\kappa^K = \delta_{a\kappa} \tag{3.9b}$$

whereas

$$A_a^K \neq B_\kappa^K \delta_a^\kappa \tag{3.10a}$$

while

$$A_a^K A_K^\kappa \neq \delta_a^\kappa \tag{3.10b}$$

These relations readily follow from inspection of Eqs. 2.10 and 3.3.

A pure plastic strain can now be defined exactly by writing an expression for length $\underset{a}{ds}$ in the plastic state (a) corresponding to that given by Eq. 2.1 for the initial reference state. In particular

$$\left(\underset{a}{ds}\right)^2 = a_{ab}\,dx^a\,dx^b \tag{3.11}$$

We can next write, similar to Eq. 2.16

$$\left(\underset{a}{ds}\right)^2 - \left(\underset{K}{ds}\right)^2 = \left(a_{ab}A^a_K A^b_L - a_{KL}\right)dx^K\,dx^L \tag{3.12}$$

The plastic strain tensor is defined as

$$e^P_{KL} = \tfrac{1}{2}\left(a_{ab}A^a_K A^b_L - a_{KL}\right) \tag{3.13}$$

and may be compared to the elastic strain counterpart given by Eq. 2.18. It is obvious that the plastic strain is always equal to its elastic strain counterpart, that is,

$$e^P_{KL} = e_{\kappa\lambda}\delta^\kappa_K\delta^\lambda_L \tag{3.14}$$

The above definition of plastic strain and its relation to elastic strain is an extremely important result which, as we shall see, has far-reaching consequences. It is unfortunate that such an important result has gone unrecognized, even in the classical texts dealing with plasticity (Freudenthal, 1958; Hill, 1950; Prager, 1959; Hoffman and Sachs, 1953).

It is now rather straightforward to describe a pure uniaxial compression shown as state (a^1) in Figure 3.1b whose elastic counterpart is the (κ^1) state of Figure 2.1c. In particular, the nonvanishing components of the plastic distortion tensor $A^{a^1}_K$ are given by

$$A^1_1 = A^3_3 = 1 \tag{3.15a}$$

while

$$A^2_2 = \tfrac{1}{2} \tag{3.15b}$$

This causes the coordinate $dx^{K=2} = 1$ in Figure 1a to revert to $dx^{a^1=2} = \frac{1}{2}$ in Figure 3.1b.

Figure 3.2a corresponds to a simple plastic shear is the discrete approximation. When the steps on the right and leftmost faces of the figure are made infinitesimally small, we obtain the continuous representation

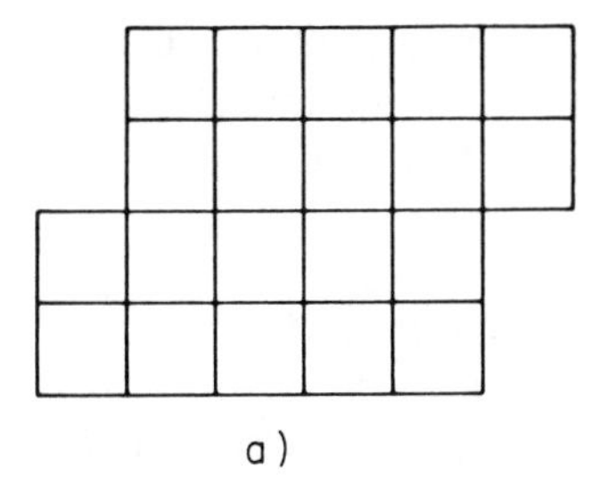

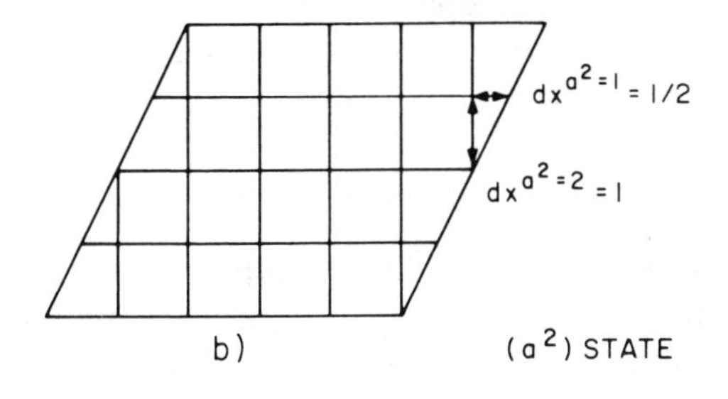

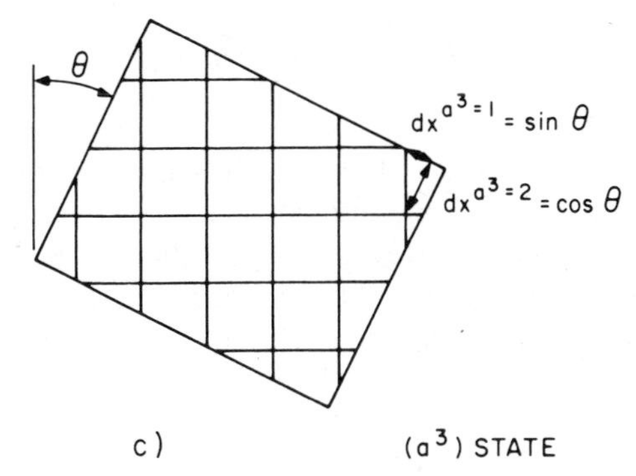

Figure 3.2 Pure plastic simple shear. *a*) Discrete representation, *b*) Continuous representation. *c*) Pure plastic rotation.

shown in Figure 3.2*b*. This is the counterpart of state (κ^2) in Figure 2.2*a* and is termed state (a^2). It may be generated from state (K) by the plastic distortion tensor A_K^a which has nonvanishing components

$$A_1^1 = A_2^2 = A_3^3 = 1 \tag{3.16a}$$

and

$$A_2^1 = \tan\theta = \tfrac{1}{2} \tag{3.16b}$$

It is apparent from these equations that the coordinate $dx^{K=2}=1$ in Figure 2.1*a* becomes the two coordinates $dx^{a=2}=1$ and $dx^{a=1}=\frac{1}{2}$ in Figure 3.2*b*. The plastic strain tensor associated with the (a^2) state is also found, similar to Eq. 3.13, to be given by

$$e^P_{KL}=\tfrac{1}{2}\left(\delta_{a^2b^2}A^{a^2}_K A^{b^2}_L-\delta_{KL}\right)=\tfrac{1}{2}\left(A^{a^2}_K A^{b^2}_L-\delta_{KL}\right) \tag{3.17}$$

which yields the same result as the elastic counterpart $e_{\kappa^2\lambda^2}$ given by Eq. 2.26.

A pure plastic rotation corresponding to the pure elastic rotation of Figure 2.2*b* is shown in Figure 3.2*c* and is denoted as state (a^3). It can be generated from the (K) state by the plastic distortion tensor $A^{a^3}_K$ where

$$A^{a^3}_K=B^L_{\kappa^3}\delta^{a^3}_L\delta^{\kappa^3}_K \tag{3.18}$$

and where $B^K_{\kappa^3}$ is given by Eq. 2.27. The distortion tensor given by the last expression causes the coordinate $dx^{K=2}=1$ to decompose into the components $dx^{a^2=1}=\sin\theta$ and $dx^{a^2=2}=\cos\theta$ shown in Figure 3.2*c*. Furthermore, the strain tensor associated with the plastic rotation is given by

$$e^P_{KL}=\tfrac{1}{2}\left(A^{a^3}_K A^{a^3}_L-\delta_{KL}\right)=0 \tag{3.19}$$

which is exactly the same result obtained for a pure elastic rotation given by Eq. 2.29. The importance of these results cannot be overstated, and it is astounding that they have gone unrecognized for so long a time.

A plastic double shear corresponding to the elastically distorted state shown in Figure 2.3*a* is illustrated in Figure 3.3*a* and denoted by (a^4). This state may be generated from the (K) state as a result of the following distortion tensor:

$$A^{a^4}_K=B^L_{\kappa^4}\delta^{a^4}_L\delta^{\kappa^4}_K \tag{3.20}$$

As a result of this distortion, the coordinate $dx^{K=2}=1$ is decomposed into $dx^{a^4=1}=\frac{1}{2}$ and $dx^{a^4=2}=1$, as indicated in Figure 3.3*a*. It is also a simple matter to show that the plastic strain tensor associated with this state is

$$e_{a^4b^4}=\delta^{\kappa^4}_{a^4}\delta^{\lambda^4}_{b^4}e_{\kappa^4\lambda^4} \tag{3.21}$$

where $e_{\kappa^4\lambda^4}$ is given by Eq. 2.32. Similar to Eq. 2.33, the plastic distortion tensor given by Eq. 3.20 can be decomposed into two separate distortions according to

$$A^{a^4}_K=A^{a^4}_{a^5}A^{a^5}_K \tag{3.22}$$

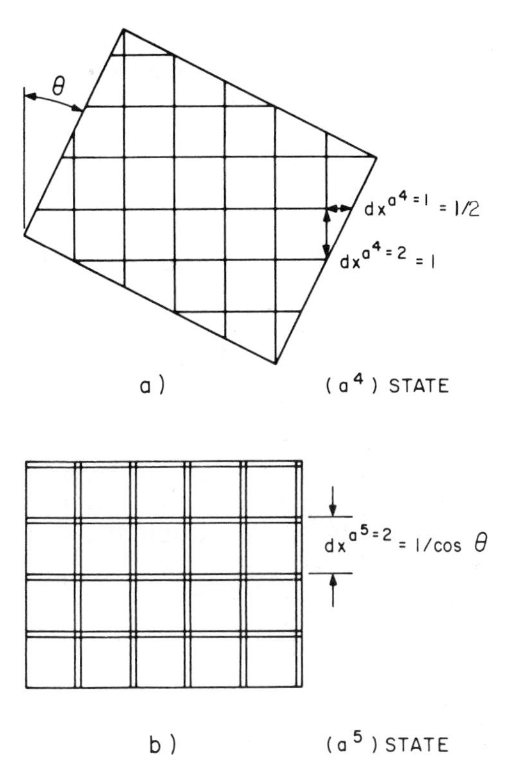

Figure 3.3 Plastically strained states of *a*) double shear, *b*) biaxial tension.

$$A_{a^5}^{a^4} = \delta_{\kappa^5}^{a^4}\delta_{a^5}^{\kappa^4}B_{\kappa^4}^{\kappa^5} \tag{3.23}$$

where $B_{\kappa^4}^{\kappa^5}$ is given by Eq. 2.34a, and is thus a simple plastic rotation. On the other hand,

$$A_K^{a^5} = \delta_L^{a^5}\delta_K^{\kappa^5}B_{\kappa^5}^L \tag{3.24}$$

where $B_{\kappa^5}^K$ is given by Eq. 2.34. Thus, $A_K^{a^5}$ gives rise to a biaxial volume expansion that generates the state (a^5) shown in Figure 3.3*b* from state (K). More specifically, this distortion changes the coordinate $dx^{K=2}=1$ to $dx^{a^5=2}=1/\cos\theta$ as indicated in Figure 3.3*b*. It is also easy to show that

$$e_{a^5b^5} = \delta_{a^5}^{a^4}\delta_{b^5}^{b^4}e_{a^4b^4} \tag{3.25}$$

Thus, analogous to the double elastic shear, the double plastic shear can be decomposed into a rigid plastic rotation plus a plastic volume expansion.

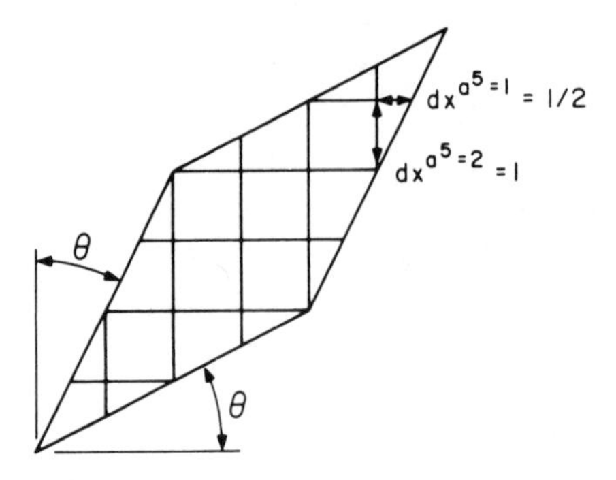

Figure 3.4 Plastically strained state of double shear, but in opposite sense.

Finally, Figure 3.4 shows the (a^6) state which is the plastic analogue of the elastically distorted (κ^6) state shown in Figure 2.4. It may be generated from the (K) state via the following plastic distortion tensor:

$$A_K^{a^6} = \delta_L^{a^6}\delta_K^{\kappa^6}B_{\kappa^6}^L \tag{3.26}$$

where $B_{\kappa^6}^K$ is given by Eq. 2.38. Furthermore, the plastic strain associated with this state is

$$e_{a^6b^6} = \delta_{a^6}^{\kappa^6}\delta_{b^6}^{\lambda^6}e_{\kappa^6\lambda^6} \tag{3.27}$$

where $e_{\kappa^6\lambda^6}$ is given by Eq. 2.40. The correspondence between the elastically distorted states and their plastically distorted analogues, which we initially set out to prove, is thus nearly complete.

The volume associated with any of the plastically distorted states considered thus far may be written as

$$V_{(a)} = \mathbf{v}_1 \times \mathbf{v}_2 \cdot \mathbf{v}_3 = \det(A_K^a) \tag{3.28}$$

where

$$\mathbf{v} = x^a\mathbf{e}_a \tag{3.29a}$$

or equivalently

$$\mathbf{v}_K = A_K^a\mathbf{e}_a \tag{3.29b}$$

Equation 3.29 corresponds to the plastic analogue of Eq. 2.41. Note that unlike the case in Eq. 2.41, the $\mathbf{e}_a$ in Eq. 3.29b are unit base vectors. Thus, analogous to Eq. 2.42, Eq. 3.28 yields

$$V_{(K)} = V_{(a^2)} = V_{(a^3)} = 1 \tag{3.30}$$

and so on.

In a manner similar to that for the (κ^2) state, a set of principal directions and eigenvalues can be associated with the (a^2) state. In particular, we may write analogous to Eq. 2.43a.

$$b_{KL} x^L = \lambda a_{KL} x^L \tag{3.31}$$

As was the case for x^{λ^2} in Eq. 2.43a, x^L in the above equation are unit lengths. The quantity b_{KL} is the metric tensor of the (a^2) state but expressed in terms of the (K) state, that is,

$$b_{KL} = A_K^{a^2} A_L^{b^2} a_{a^2 b^2} \tag{3.32}$$

where

$$a_{a^2 b^2} = \delta_{a^2 b^2} \tag{3.33}$$

so that Eq. 3.31 can be rewritten as

$$A_K^{a^2} A_L^{a^2} x^L = \lambda \delta_{KL} x^L \tag{3.34}$$

The above equation has a solution when

$$\det\left(A_K^{a^2} A_L^{a^2} - \lambda \delta_{KL}\right) = 0 \tag{3.35}$$

which yields

$$b_{K^D L^D} = A_{K^D}^{a^D} A_{L^D}^{a^D} = \delta_{K^D}^{\kappa^D} \delta_{L^D}^{\lambda^D} g_{\kappa^D \lambda^D} \tag{3.36}$$

where $g_{\kappa^D \lambda^D}$ is the same as that given by Eq. 2.46. In particular, the eigenvalues associated with simple plastic shear became exactly identical to those obtained for the corresponding elastic distortion. Again, as was the case for an elastic distortion, these results can be described in terms of plastic distortion of a unit circle into an ellipsoid by the plastic distortion tensor $A_K^{a^2}$ and is illustrated in Figure 3.5. The roots $\underset{1}{\lambda}$ and $\underset{2}{\lambda}$ correspond to the following lengths

$$(ds)^2 = \lambda_{11} x^{K^D=1} x^{K^D=1} = x^{a^D=1} x^{a^D=1} \tag{3.37a}$$

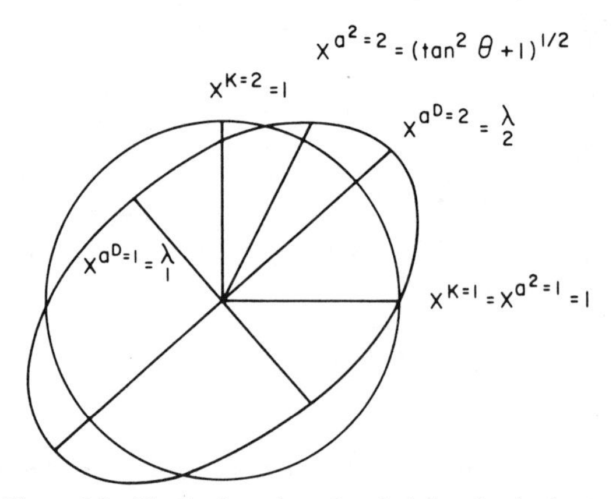

Figure 3.5 Plastic distortion of a circle by simple shear.

and

$$(ds)^2 = \lambda_{22} x^{K^D=2} x^{K^D=2} = x^{a^D=2} x^{a^D=2} \tag{3.37b}$$

which are indicated in Figure 3.5. The eigenvalues given by Eq. 3.36 can in turn be substituted into Eq. 3.34 to obtain a set of eigenvectors that are identical to those given by Eq. 2.53 for the corresponding elastic distortion.

In concluding this section, we formulate a very important principle: namely, that it is possible to relate the plastic and elastic distortions by means of a coordinate transformation. As an example, consider the (a) state of Figure 3.1a. It may be related to the (κ) state of Figure 2.1b as follows:

$$A_K^\kappa = C_a^\kappa A_K^a = \delta_K^\kappa \tag{3.38}$$

where

$$C_a^\kappa = \delta_a^K \delta_\lambda^\kappa B_K^\lambda \tag{3.39}$$

Similarly

$$B_K^\kappa = C_a^\kappa B_K^a = C_a^\kappa \delta_K^a \tag{3.40}$$

Similar relations can be written between the remaining plastic states (a^1), (a^2), (a^3), (a^4), (a^5), and (a^6) and their corresponding elastic states (κ^1), (κ^2), (κ^3), (κ^4), (κ^5), and (κ^6), respectively.

REVIEW

The primary goal of the present chapter has been to formulate a suitable concept of plastic distortion. In particular, a plastically strained state (a) may be generated from an unstrained reference state (K) by means of a plastic distortion tensor A_K^a. This distortion tensor connects the coordinates dx^a and dx^K between the strained and unstrained states, respectively, and it is these which change in response to a plastic strain. In contrast to an elastic distortion discussed in the previous chapter, the metric tensor g_{KL} is unaffected by a plastic distortion, that is one may say that it is dragged along by the distortion. This is also equivalent to writing the elastic distortion tensor B_K^a for a plastic distortion as δ_K^a that in effect says that the elastic distortion vanishes. Similarly, for a pure elastic distortion, the plastic distortion tensor A_K^κ can be written as δ_K^κ which corresponds to a vanishing of the plastic distortion. With these definitions, it is possible to show that any possible pure elastic distortion has associated with it a corresponding pure plastic distortion since we can always write $A_K^a = B_K^\kappa \delta_\kappa^a$. This leads to some interesting results when elastic and plastic rotations are considered. In particular, Figure 2.2*b* corresponds to a pure elastic rotation, which is also analagous to a rigid rotation, whereas Figure 3.2*c* corresponds to a pure plastic rotation. Note that the base vectors remain unrotated in the latter figure, that is, are dragged. As in the case of an elastic strain, the distortion tensor which yields the plastic strain can be used to define a plastic strain tensor. The elastic and plastic strain tensors are identical for the same types of elastic and plastic strains given by $A_K^a = B_K^\kappa \delta_\kappa^a$.

In summary, we are led to the extremely important result which says that whereas a pure elastic distortion involves a change in the metric tensor and a corresponding dragging (constancy) of the coordinates, a pure plastic distortion involves a change in the coordinates and a corresponding dragged or unaltered metric tensor. As is shown in the following chapter, the precise definition of a plastic distortion is essential to the proper description of the dislocated state.

CHAPTER 4

Dislocations

The plastic states discussed in the previous chapter may be generated from the initial undeformed reference state by means of dislocations. When these dislocations terminate within the interior of the body, the plastic distortion may be viewed as imperfect or incomplete. Furthermore, these dislocations are associated with elastic distortions. Our aim in this chapter is to provide as complete description of this dislocated state as possible. The geometrical methods are quite general and apply to arbitrarily large distortions.

4.1 DISTORTIONS

The question that naturally arises now is whether the plastically distorted states considered in the previous section contain dislocations. This is best answered by considering the prismatic dislocation configuration illustrated in Figure 4.1*a* that consists of a pair of edge dislocations of opposite sign. After the leftmost dislocation has moved to the surface of the crystal, the configuration shown in Figure 4.1*b* obtains. It is apparent that such dislocation formation ultimately leads to the (a) state configuration shown in Figure 3.1*a*. These dislocated states are denoted by lowercase Latin letters that occupy the center of the alphabet, that is, k, l, m, and so on. It is also obvious that the dislocated states in Figure 4.1 have associated with them both plastic as well as elastic strains. The dislocated (k) state configuration can be generated from the (K) state by means of the distortion tensor A_K^k that relates the coordinates in the two states as follows:

$$dx^k = A_K^k dx^K \tag{4.1}$$

In the case of Figure 4.1*b*, the nonvanishing components of A_K^k are

$$A_1^1 = A_3^3 = 1 \tag{4.2a}$$

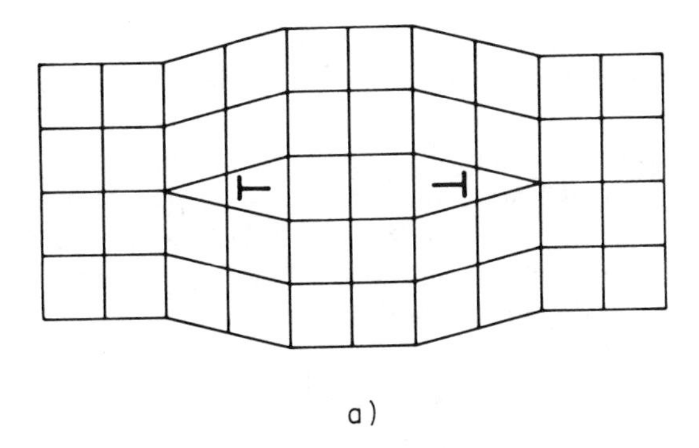

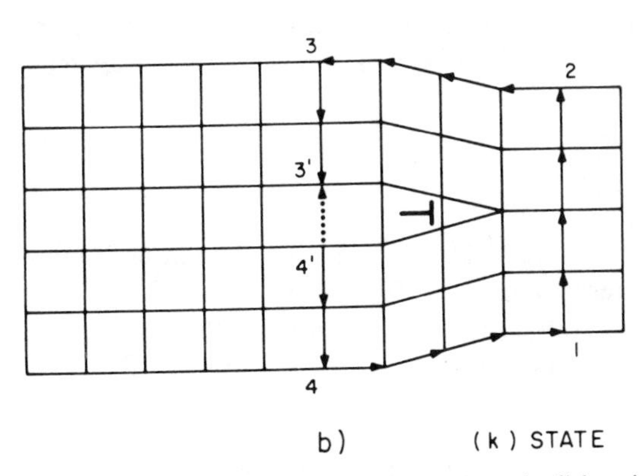

Figure 4.1 *a*) Uniaxial plastic elongation arising from prismatic dislocation formation. *b*) Same as *a*), but after leftmost dislocation has moved to surface of body.

while

$$A_2^2 = H(-x^1)\left[\delta(x^2)+1\right] + H(+x^1) \tag{4.2b}$$

where $H(-x^1)$ and $H(+x^1)$ are Heaviside functions defined such that

$$H(-x^1) = \begin{cases} 0 & \text{if } x^1 > 0 \\ 1 & \text{if } x^1 < 0 \end{cases} \tag{4.3a}$$

while

$$H(+x^1) = \begin{cases} 0 & \text{if } x^1 < 0 \\ 1 & \text{if } x^1 > 0 \end{cases} \tag{4.3b}$$

The quantity $\delta(x^2)$ is the Dirac delta function defined such that

$$\delta(x^2) = 0 \tag{4.4}$$

for $x^2 \neq 0$. The treatment of Heaviside and Dirac delta functions can be found in a number of works (Kunin, 1965; Gel' Fond and Shilov, 1964; deWit, 1973). The coordinates x^1 and x^2 in Eq. 4.2b are measured with respect to the position of the dislocation whose origin is taken as zero. In words, Eqs. 4.1 and 4.2 state that the (k) state has no plastic distortion associated with it for $x^1 > 0$ and for $x^2 \neq 0$. On the other hand, for $x^1 < 0$ and $x^2 = 0$, $dx^{K=2} = 1$ transforms to $dx^{k=2} = 2$. This arises from the fact that $\delta(x^2)$ has the following property:

$$\int_{-\infty}^{+\infty} \delta(x^2)\, dx^2 = 1 \tag{4.5}$$

Since the dislocated (k) state also contains elastic distortions, we must also define an elastic distortion B_k^K that relates the base vectors in the two states as follows:

$$\mathbf{e}_k = B_k^K \mathbf{e}_K \tag{4.6}$$

In the case of Figure 4.1b, B_k^K is some suitable function of position measured with respect to the dislocation.

A strain tensor can next be formulated with respect to the (k) state by first writing for the measure of distance in this state

$$\left(\underset{k}{ds}\right)^2 = g_{kl}\, dx^k\, dx^l \tag{4.7a}$$

where g_{kl} is the metric tensor associated with the (k) state or

$$g_{kl} = \mathbf{e}_k \cdot \mathbf{e}_l = B_k^K B_l^K \tag{4.7b}$$

In view of Eqs. 4.7 and 2.1, we can now write

$$\left(\underset{k}{ds}\right)^2 - \left(\underset{K}{ds}\right)^2 = \left(g_{kl}\, dx^k\, dx^l - a_{KL}\, dx^K\, dx^L \right) \tag{4.8}$$

Eq. 4.1 now allows us to write the last equation as

$$\left(\underset{k}{ds}\right)^2 - \left(\underset{K}{ds}\right)^2 = \left(g_{kl} A_K^k A_L^l - a_{KL} \right) dx^K\, dx^L \tag{4.9}$$

where the strain tensor is defined as

$$\underset{D}{e}_{KL} = \tfrac{1}{2}\left(g_{kl} A_K^k A_L^l - \delta_{KL} \right) \tag{4.10}$$

where the subscript D denotes dislocation. Note that for the elastically strained regions about the dislocation in Figure 4.1b, $A_L^k = \delta_L^k$, so that Eq. 4.10 reduces to the same expression given for Eq. 2.18 for pure elastic strain. On the other hand, for the plastically strained region associated with the dislocation, that is, $x^1 \ll 0$ and $x^2 = 0$, $g_{kl} = \delta_{kl}$, so that Eq. 4.10 reduces to the same expression given by Eq. 3.13 for pure plastic strain.

If a uniform array of dislocations such as those shown in Figure 4.1b is added to the leftmost side of the (K) state crystal, the (k') state configuration shown in Figure 4.2 obtains. The dislocations therein give rise to a plastic strain within the leftmost portion of the body that is identical to the (a) state configuration shown in Figure 3.1a. The dislocation tensor that accomplishes this deformation is given by $A_K^{k'}$ where the nonvanishing components are

$$A_1^1 = A_3^3 = 1 \tag{4.11a}$$

while

$$A_2^2 = \{2H(-x^1)\}_1 + \{H(+x^1)\}_2 \tag{4.11b}$$

where x^1 is measured from the vertical wall of dislocations. Note that because of the uniform distortion in the x^2 direction, $\delta(x^2)$, as given in Eq. 4.2b, is no longer necessary in the last expression. Note also that the

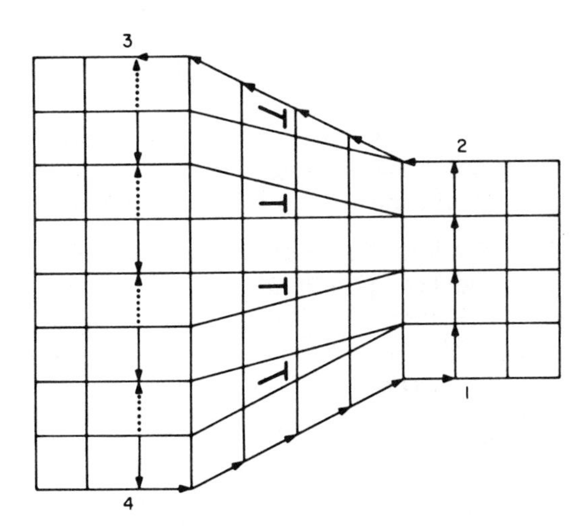

Figure 4.2 Uniform uniaxial plastic elongation arising from dislocation formation.

dislocation array essentially divides the crystal into two regions, that is, a leftmost plastically distorted region, and a rightmost nonplastically distorted region. This division into two regions is emphasized by the use of curly brackets in Eq. 4.11b. Such a notation is used extensively throughout the remaining sections. Again, because of the nonvanishing elastic distortions associated with state (k'), we must write

$$g_{k'l'} = B_{k'}^{K} B_{l'}^{K} \tag{4.12}$$

whereas the corresponding strain tensor becomes

$$\underset{D}{e_{KL}} = \tfrac{1}{2}\left(g_{k'l'} A_K^{k'} A_L^{l'} - \delta_{KL} \right) \tag{4.13}$$

When an edge dislocation dipole of opposite sense to that shown in Figure 4.1*a* is nucleated within the perfect (K) state crystal, the configuration shown in Figure 4.3*a* obtains. When the leftmost dislocation is allowed to move to the surface of the crystal, the (k^1) state configuration illustrated in Figure 4.3*b* obtains. The nonvanishing components of the distortion tensor $A_K^{k^1}$ giving rise to the dislocated (k^1) state may be written as

$$A_1^1 = A_3^3 = 1 \tag{4.14a}$$

and

$$A_2^2 = H(-x^1)\left[-\delta(x^2) + 1 \right] + H(+x^1) \tag{4.14b}$$

Except for the factor of $1/2$, the last expression is the same as that given by Eq. 4.2b which generated the (k) state from state (K). A metric $g_{k^1l^1}$ and strain tensor $\underset{D}{e_{KL}}$ can also be written for the (k^1) state similar to that done for state (k). If a uniform array of dislocations such as those shown in Figure 4.3*b* is introduced into the (K) state crystal, the ($k^{1'}$) state configuration shown in Figure 4.3*c* obtains. Such a dislocation configuration is closely related to the (a^1) state configuration shown in Figure 3.1*b* and can in fact be used to generate this state. The nonvanishing components of the distortion tensor $A_K^{k^{1'}}$ becomes

$$A_1^1 = A_3^3 = 1 \tag{4.15}$$

and

$$A_2^2 = \left\{ \tfrac{1}{2} H(-x^1) \right\}_1 + \left\{ H(+x^1) \right\}_2 \tag{4.16}$$

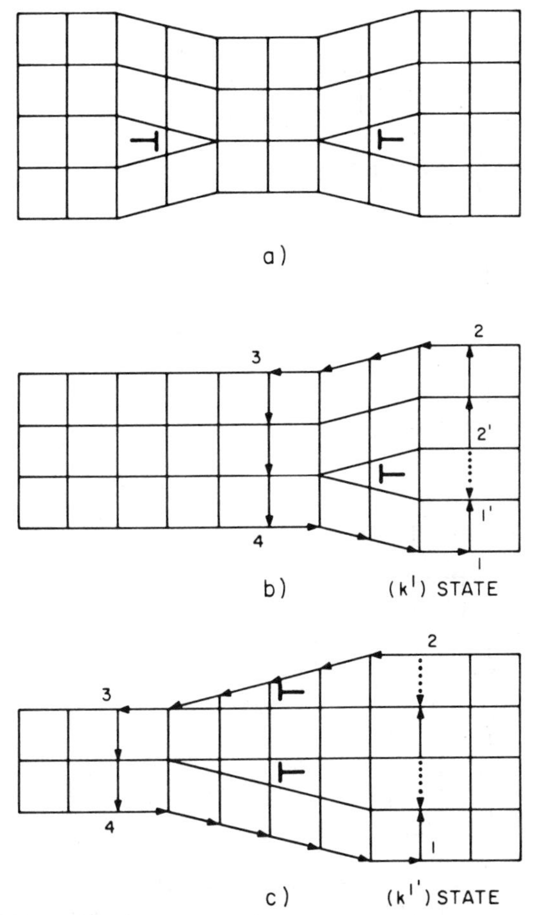

Figure 4.3 *a*) Uniaxial plastic compression arising from dislocation dipole formation. *b*) Same as *a*), but after leftmost dislocation has moved to surface of body. *c*) Distortion similar to that in *b*), but with a uniform distribution of dislocations.

where the similarity with Eq. 4.11 is obvious. Again, the metric and strain tensors associated with state $(k^{1'})$ are readily obtained.

When an edge type glide type dislocation dipole is nucleated within a perfect crystal, the configuration shown in Figure 4.4*a* obtains. Upon glide of the leftmost dislocation of this pair to the surface of the body, the (k^2) state crystal shown in Figure 4.4*b* is generated. The nonzero components of the distortion tensor $A_K^{k^2}$ giving rise to the (k^2) state configuration may be written as

$$A_2^2 = A_3^3 = 1 \tag{4.17a}$$

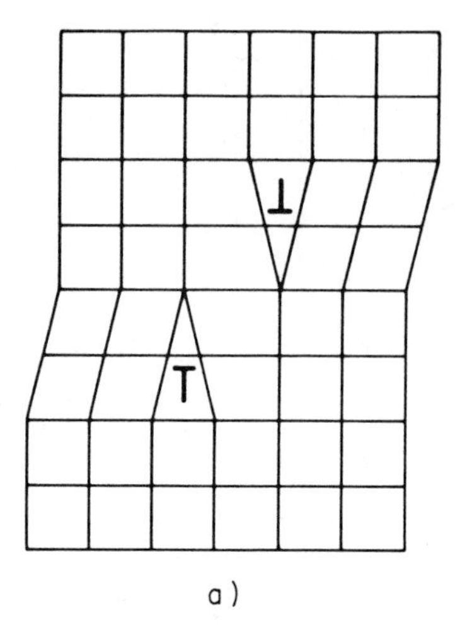

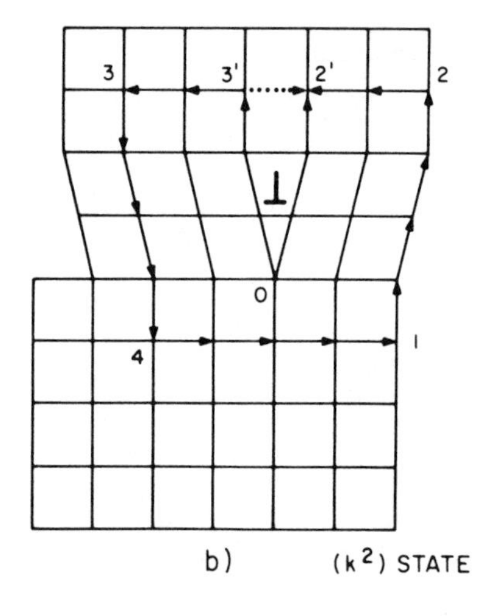

Figure 4.4 *a*) Plastic shear resulting from glide dislocation dipole formation. *b*) Same as *a*), but after leftmost dislocation has moved to surface of body.

and

$$A_1^1 = H(+x^2)\left[\delta(x^1)+1\right] + H(-x^2) \tag{4.17b}$$

When a uniform vertical distribution of dislocations such as that shown in Figure 4.4*b* is introduced into the crystal, the $(k^{2'})$ state configuration shown in Figure 4.5 obtains. The $(k^{2'})$ state is obviously the precursor of the (a^2) state shown in Figure 3.2*a*. Its distortion tensors are given by $A_K^{k^{2'}}$ that has nonzero components

$$A_1^1 = A_2^2 = A_3^3 = 1 \tag{4.18a}$$

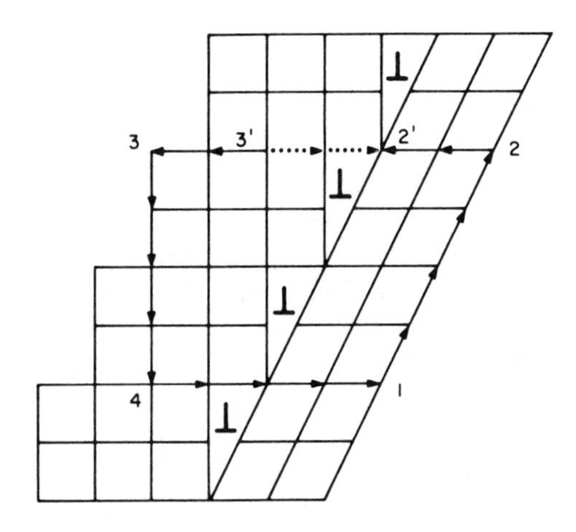

Figure 4.5 Plastic distortion similar to that shown in Figure 4.4*b* but with a uniform distribution of dislocations.

and

$$A_2^1 = \{\tan\theta H(-x^1)\}_1 \tag{4.18b}$$

where $\tan\theta$ was chosen as $1/2$. Again, the $(k^{2'})$ and (k^2) states possess metric and strain tensors that are of the same form as those given by Eqs. 4.12 and 4.13. It is also possible to find dislocated states (k^3), (k^4), (k^5), and (k^6) as well as their more continuous analogues $(k^{3'})$, $(k^{4'})$, $(k^{5'})$, and $(k^{6'})$ corresponding to the states (a^3), (a^4), (a^5), and (a^6). Although somewhat more complex than the dislocated states already considered, they nevertheless consist of the same basic dislocation arrays and are not considered further.

We have seen in the previous two sections that a given state of elastic or plastic strain can be represented in different ways depending upon the orientation of the orthogonal Cartesian coordinate axes. However, since the strain tensor $\mathbf{e}$ is an invariant, it cannot depend upon the coordinate axes, as can be seen by the following relation (Fung, 1965; Coburn, 1970)

$$\mathbf{e} = e_{\kappa\lambda}\mathbf{e}^{\kappa}\mathbf{e}^{\lambda} = e^{\kappa\lambda}\mathbf{e}_{\kappa}\mathbf{e}_{\lambda} \tag{4.19}$$

where $\mathbf{e}^{\kappa}$ and $e^{\kappa\lambda}$ are the contravariant components of the base vectors and strain tensors in the coordinate system corresponding to state (κ). Thus, whereas the components of the strain tensor $e_{\kappa\lambda}$ vary, depending upon the coordinate system, the strain tensor itself $\mathbf{e}$ is an invariant, and thus is independent of the coordinate system.

The question that now arises is whether the dislocation configurations considered in this section can also be represented in alternate ways, that is, in terms of different coordinate systems. This can be most readily answered by considering the dislocation configuration shown in Figure 4.6. It is clear that this is simply an alternate representation of the configuration shown in Figure 4.2. In particular, it is simply the representation of the (k') state in a different coordinate system. This can be made clearer by considering the (K^c) state shown in Figure 4.7*a*. This is simply the (K) state of Figure 2.1*a* in a different coordinate system. The superscript c in (K^c) is used to denote a different coordinate system. The $(K)\rightarrow(K^c)$ transformation can now be written as

$$C_K^{K^c} = \begin{bmatrix} \sqrt{2}\cos\theta & -\sqrt{2}\sin\theta & 0 \\ \sqrt{2}\sin\theta & \sqrt{2}\cos\theta & 0 \\ 0 & 0 & 1 \end{bmatrix} \qquad (4.20)$$

where θ in Figure 4.7*a* has been chosen as 45°. The kernel letter c is used to denote coordinate transformations. It also follows that

$$C_K^{K^c} C_{K^c}^L = \delta_K^L \qquad (4.21)$$

Also, unlike the elastic and plastic distortions, B_κ^K and A_K^a, respectively, given by Eqs. 2.8 and 3.3, respectively, the coordinate transformation $C_K^{K^c}$ connects the base vectors, as well as the coordinates in the two frames. In

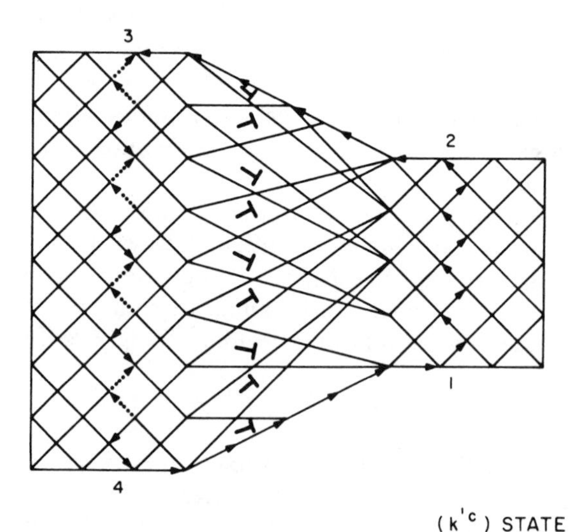

Figure 4.6 Alternate representation of the dislocation configuration shown in Figure 4.2.

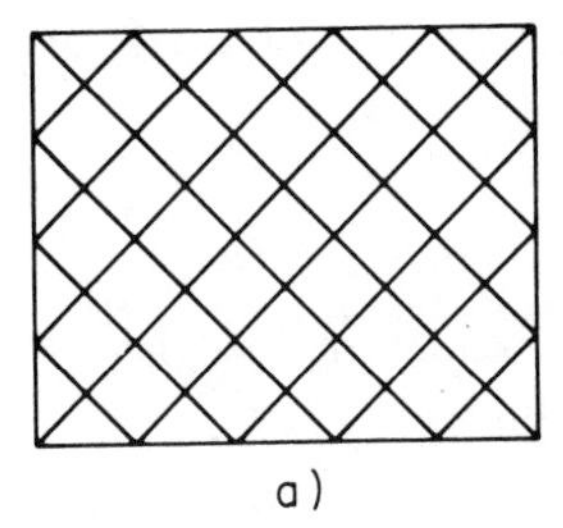

Figure 4.7 Alternate representation *a*) of the (K) state and *b*) of the (k) states shown in Figures 2.1*a* and 3.1*a*, respectively.

particular

$$\mathbf{e}_{K^c} = C^{K}_{K^c}\mathbf{e}_K \tag{4.22a}$$

and

$$\mathbf{e}_K = C^{K^c}_{K}\mathbf{e}_{K^c} \tag{4.22b}$$

while

$$dx^{K^c} = C^{K^c}_{K}\, dx^K \tag{4.23a}$$

and

$$dx^K = C^{K}_{K^c}\, dx^{K^c} \tag{4.23b}$$

Again, the metric tensor in the transformed system (K^c) is of the form given by Eq. 2.11, that is,

$$g_{K^cL^c} = C^K_{K^c} C^K_{L^c} \tag{4.24}$$

If we now attempt to use the metric in the above equation to define a strain tensor of the form given by Eq. 2.17, we obtain

$$e_{K^cL^c} = \tfrac{1}{2}\left(g_{K^cL^c} - \delta_{KL} C^K_{K^c} C^L_{L^c} \right) = 0 \tag{4.25}$$

This is exactly as it should be since there can be no strain associated merely with a change in coordinates.

The (a^c) state configuration shown in Figure 4.7*b* is just an alternate representation of the (a) state configuration shown in Figure 3.1*a*, but in terms of the same coordinate system shown in Figure 4.7*a*. In particular, it may be generated from the (K) state by the following distortion:

$$A^{a^c}_K = A^a_K C^{a^c}_a \tag{4.26}$$

where A^a_K is given by Eq. 3.3, while

$$C^{a^c}_a = \delta^K_a \delta^{a^c}_{K^c} C^{K^c}_K \tag{4.27}$$

where $C^{K^c}_K$ is given by Eq. 4.20. In expanded form, Eq. 4.26 becomes

$$A^{a^c}_K = \begin{pmatrix} 2\sqrt{2}\cos\theta & -\sqrt{2}\sin\theta & 0 \\ 2\sqrt{2}\sin\theta & 2\sqrt{2}\cos\theta & 0 \\ 0 & 0 & 1 \end{pmatrix} \tag{4.28}$$

It is now important to note that the above distortion connects the coordinates as follows:

$$dx^{a^c} = A^{a^c}_K \, dx^K \tag{4.29a}$$

and

$$dx^K = A^K_{a^c} \, dx^{a^c} \tag{4.29b}$$

These distortions, however, do not connect the base vectors since the base vectors must be dragged with respect to any plastic deformations. This means that we must write

$$\mathbf{e}_{a^c} = B^a_{a^c} \mathbf{e}_a \tag{4.30a}$$

and

$$\mathbf{e}_a = B_a^{a^c} \mathbf{e}_{a^c} \tag{4.30b}$$

where

$$B_a^{a^c} = \delta_a^b \delta_{b^c}^{a^c} C_b^{b^c} \tag{4.31}$$

where $C_b^{b^c}$ is given by Eq. 4.27. The plastic strain associated with the $(K) \to (a^c)$ distortion may be written, similar to Eq. 3.13, as

$$e_{KL}^P = \tfrac{1}{2}\left(g_{a^c b^c} A_K^{a^c} A_L^{b^c} - a_{KL} \right) \tag{4.32}$$

also

$$g_{a^c b^c} = A_{a^c}^K A_{b^c}^K = C_{a^c}^a A_a^K C_{b^c}^b A_b^K \tag{4.33}$$

where the last expression follows from Eq. 4.26. Upon substitution of the last equation into Eq. 4.32, we obtain

$$e_{KL}^P = \tfrac{1}{2}\left(A_a^K A_b^K A_K^{a^c} C_{a^c}^a A_L^{b^c} C_{b^c}^b - a_{KL} \right) = \tfrac{1}{2}\left(a_{ab} A_K^a A_L^b - a_{KL} \right) \tag{4.34}$$

This is exactly the same result as that given by Eq. 3.13 and shows that the distortions $(K) \to (a)$ and $(K) \to (a^c)$ have associated with them the same plastic strains, that is, the strain cannot depend upon the coordinate system associated with the final state.

We are finally in a position to write the distortion tensor associated with the $(K) \to (k'^c)$ distortion that leads to the dislocated state shown in Figure 4.6. Specifically,

$$A_K^{k'^c} = \left\{ \underset{1}{A_K^{k'^c}} H(-x^1) \right\}_1 + \left\{ \underset{2}{A_K^{k'^c}} H(+x^1) \right\}_2 \tag{4.35}$$

where

$$\underset{1}{A_K^{k'^c}} \equiv A_K^{a^c} \tag{4.36a}$$

as given by Eq. 4.28 while

$$\underset{2}{A_K^{k'^c}} \equiv C_K^{K^c} \tag{4.36b}$$

as given by Eq. 4.20. The coordinates x^K in the Heaviside function are measured with respect to the (K) state.

4.2 BURGERS VECTOR

Perhaps nothing has been more confusing in the literature relating to dislocation theory than the concept of a Burgers circuit (Hirth and Lothe, 1968; Marcinkowski, 1978c; Frank 1951). Part, if not all, of this confusion seems to have resulted from the failure to understand the exact mathematical significance of such a reference circuit. Consider first of all the reference circuit 1–2–3–4–1 shown by the dotted arrows in Figure 2.1*a* in the (K) state. This same circuit is distorted into that given by 1–2–3–3′–4′–4–1 in Figure 4.1*b* for the (k) state. The Burgers vector or closure failure is given by the dotted line 4′–3′. Mathematically, this closure failure may be written in terms of the following line integral (Kröner, 1959)

$$b^k = -\oint A_K^k \, dx^K \tag{4.37}$$

This is simply the negative of the line integral of dx^k given by Eq. 4.1. Equation 4.37 can be expanded to yield

$$b^{k=2} = -\int_1^2 A_2^2 dx^2 - \int_2^3 A_1^2 dx^1 - \int_3^4 A_2^2 dx^2 - \int_4^1 A_1^2 dx^1 \tag{4.38a}$$

that in terms of the distortions given by Eq. 4.2 gives

$$b^{k=2} = -\int_1^2 A_2^2 dx^2 - \int_3^4 A_2^2 dx^2 \tag{4.38b}$$

or

$$b^{k=2} = -\int_1^2 dx^2 - \int_3^4 [1+\delta(x^2)]\, dx^2 \tag{4.38c}$$

or still more simply

$$b^{k=2} = \{-4\} + \{4+1\} = 1 \equiv \underset{4'-3'}{\Delta x^2} \tag{4.38d}$$

The quantity $\underset{4'-3'}{\Delta x^2}$ corresponds to the dotted closure failure in Figure 4.1*b*. It is important to note that $b^{k=2}$ is only a component measured in the (k) system. The closure vector must be written as

$$\mathbf{b} = b^k \mathbf{e}_k \tag{4.39}$$

where $\mathbf{e}_k$ are the base vectors in the (k) system. For the case shown in Figure 4.1*b*, Eq. 4.39 becomes

$$\mathbf{b} = b^2 \mathbf{e}_2 = \mathbf{e}_2 \tag{4.40}$$

It is noted that as the portion of the circuit 3–3′–4′–4 in Figure 4.1*b* is brought closer to the dislocation, **b** becomes smaller and smaller.

For the uniform distribution of dislocations associated with the (k') state of Figure 4.2, the reference circuit 1–2–3–4–1 contains four dislocations that are shown by the dotted arrows along line 3–4. Quantitatively, we may write, similar to Eq. 4.37

$$b^{k'} = -\oint A_K^{k'} dx^K \tag{4.41}$$

In view of Eq. 4.11, this becomes

$$b^{k'=2} = -\int_1^2 dx^2 - 2\int_3^4 dx^2 = -\{4\}_2 + \{8\}_1 = 4 \tag{4.42}$$

that corresponds to the four dotted arrows shown in Figure 4.2. The closure failure associated with the (k') state could also have been written in terms of the elastic distortion tensor as follows:

$$\mathbf{b}_{k'} = -\oint B_{k'}^K d\mathbf{e}_K \tag{4.43}$$

Turning to the (k^1) state of Figure 4.3*b*, the reference circuit 1–2–3–4–1 in Figure 2.1*a* is seen to be distorted into that given by 1–1′–2′–2–3–4–1, which contains the closure failure 2′–1′. We can express this closure failure in terms of the following line integral:

$$b^{k^1} = -\oint A_K^{k^1} dx^K \tag{4.44}$$

which, in view of Eq. 4.14, becomes

$$b^{k^1=2} = -\int_1^2 dx^2 - \int_3^4 [-\delta(x^2)+1]\, dx^2 \tag{4.45a}$$

or more simply

$$b^{k^1=2} = \{-4\} - \{1-4\} = -1 \equiv \underset{2'-1'}{\Delta x^2} \tag{4.45b}$$

that is just the closure failure 2′–1′ shown dotted in Figure 4.3*b*.

In the case of the continuous distribution of dislocations that lead to state ($k^{1'}$) in Figure 4.3*c*, the reference circuit 1–2–3–4–1 contains a pair of dislocations shown dotted that lie along the line segment 2–1. In this case, the Burgers vector may be written as

$$b^{k^{1'}} = -\oint A_K^{k^{1'}} dx^K \tag{4.46}$$

that in conjunction with Eq. 4.16 yields

$$b^{k^{1'}=2} = -\int_1^2 dx^2 - \tfrac{1}{2}\int_3^4 dx^2 = \{-4\}_2 + \{2\}_1 = \{-2\}_2 \qquad (4.47)$$

that corresponds to the two dotted arrows shown in Figure 4.3*c*.

Turning now to the (k^2) state shown in Figure 4.4*b*, the Burgers circuit 1–2–2′–3′–3–4–1 contains the closure failure 3′–2′ that may be obtained from the following line integral:

$$b^{k^2} = -\oint A_K^{k^2}\, dx^K \qquad (4.48)$$

which, in view of Eq. 4.17, becomes

$$b^{k^2=1} = -\int_2^3 A_1^1\, dx^1 - \int_4^1 A_1^1\, dx^1 \qquad (4.49a)$$

which reduces to

$$b^{k^2=1} = -\int_2^3 [1+\delta(x^1)]\, dx^1 - \int_4^1 dx^1 \qquad (4.49b)$$

or finally

$$b^{k^2=1} = \{4+1\} + \{-4\} = +1 \equiv \underset{3'-2'}{\Delta x^1} \qquad (4.49c)$$

which is just the closure failure shown dotted in Figure 4.4*b*.

In terms of the continuous distribution of dislocations that gives rise to the $(k^{2'})$ state shown in Figure 4.5, the Burgers circuit 1–2–2′–3′–3–4–1 contains the closure failure 3′–2′. In terms of a line integral, we may write

$$b^{k^{2'}} = -\oint A_K^{k^{2'}}\, dx^K \qquad (4.50)$$

which, together with Eq. 4.18 yields

$$b^{k^{2'}=1} = -\int_1^2 A_2^1\, dx^2 - \int_2^3 A_1^1\, dx^1 - \int_3^4 A_2^1\, dx^2 - \int_4^1 A_1^1\, dx^1 \qquad (4.51a)$$

or in expanded form

$$b^{k^{2'}=1} = -\int_2^3 dx^1 - \tan\theta \int_3^4 dx^2 - \int_4^1 dx^1 \qquad (4.51b)$$

which reduces to

$$b^{k^{2'}=1} = \{+4\} + \{4\tan\theta\} + \{-4\} = 2 \equiv \underset{3'-2'}{\Delta x^1} \qquad (4.51c)$$

which corresponds to the dotted closure failure shown in Figure 4.5.

We can now turn our attention to the (k'^c) state shown in Figure 4.6 where it is noted that the Burgers circuit 1–2–3–4–1 contains closure failures shown dotted along the line 3–4. For convenience those portions of the circuit along line segments 2–3 and 4–1 have been drawn in terms of the (k') state shown in Figure 4.2. However, along the line segments 1–2 and 3–4, the reference circuit is in terms of the new (k'^c) state coordinates. The Burgers vector may now be written as

$$b^{k'^c} = -\oint A_K^{k'^c}\, dx^K \tag{4.52}$$

When used with Eq. 4.35, the above expression yields

$$b^{k'^c=1} = -\int_1^2 A_2^1 dx^2 - \int_3^4 A_2^1 dx^2 \tag{4.53a}$$

and

$$b^{k'^c=2} = -\int_1^2 A_2^2 dx^2 - \int_3^4 A_2^2 dx^2 \tag{4.53b}$$

which reduces to

$$b^{k'^c=1} = -\sqrt{2}\,\sin\theta \int_1^2 dx^2 - 2\sqrt{2}\,\sin\theta \int_3^4 dx^2 \tag{4.54a}$$

and

$$b^{k'^c=2} = -\sqrt{2}\,\cos\theta \int_1^2 dx^2 - 2\sqrt{2}\,\cos\theta \int_3^4 dx^2 \tag{4.54b}$$

or finally

$$b^{k'^c=1} = \{-4\} + \{8\} = 4 \tag{4.55a}$$

and

$$b^{k'^c=2} = \{-4\} + \{8\} = 4 \tag{4.55b}$$

In terms of Figure 4.6, the first of the above equations corresponds to the four dotted rightmost pointing arrows, while the second equation corresponds to the four dotted leftmost pointing arrows.

Thus we see that, as in the case of pure elastic and pure plastic distortions, any given dislocation configuration can be represented in an indefinite number of coordinate representations. Another facet of this problem is that since the Burgers vector associated with a given dislocation array is an invariant or tensor quantity, we may write

$$\mathbf{b} = b^a \mathbf{e}_a = b^{k'^c} \mathbf{e}_{k'^c} \tag{4.56}$$

It is really the above equation that enables us to write the decomposition of a dislocation as

$$\mathbf{b} \rightarrow \mathbf{b}_1 + \mathbf{b}_2 \tag{4.57a}$$

or equivalently

$$\mathbf{b} \rightarrow b^1 \mathbf{e}_1 + b^2 \mathbf{e}_2 \tag{4.57b}$$

More insight can be obtained with respect to the nature of the Burgers circuit by converting the line integral of Eq. 4.37 into a surface integral by means of Stokes' theorem as follows (Schouten, 1954; Schouten, 1951):

$$b^k = -\oint A_K^k dx^K = -\int_s \partial_{[L} A_{K]}^k dF^{LK} \tag{4.58}$$

where the bracketed term signifies that only the antisymmetric part is to be considered, that is,

$$\partial_{[L} A_{K]}^k = \tfrac{1}{2}\left[\partial_L A_K^k - \partial_K A_L^k\right] \tag{4.59}$$

Rather than carry out the surface integration given by Eq. 4.58 with respect to the (K) state, we can integrate with respect to the (k) state by first writing

$$b^k = -\int_s A_l^L A_m^K \partial_{[L} A_{K]}^k dF^{lm} \tag{4.60}$$

The above integral could also be rewritten as

$$b^k = -\int_s S_{lm}^{\cdot\cdot k} dF^{lm} \tag{4.61}$$

where the quantity $S_{lm}^{\cdot\cdot k}$ is an asymmetric tensor termed the torsion tensor which may be written as

$$S_{lm}^{\cdot\cdot k} = A_l^L A_m^K \partial_{[L} A_{K]}^k \tag{4.62}$$

The only nonvanishing component of $S_{lm}^{\cdot\cdot k}$ associated with the (k) state shown in Figure 4.1*b* may be written as

$$S_{12}^{\cdot\cdot 2}=\tfrac{1}{2}\bar{A}_1^1\bar{A}_2^2\partial_1 A_2^2 \tag{4.63}$$

where the barred quantities, that is, $\bar{A}_1^1$ is used to denote the inverse distortions. In view of Eq. 4.2, Eq. 4.63 becomes

$$S_{12}^{\cdot\cdot 2}=-\tfrac{1}{2}[\delta(x^2)+1]\delta(x^1)+\tfrac{1}{2}\delta(x^1)=-\tfrac{1}{2}\delta(x^2)\delta(x^1) \tag{4.64}$$

where use has been made of the following relations (de Wit, 1973)

$$\partial_1 H(x^1)=\delta(x^1) \tag{4.65a}$$

and

$$\partial_1 H(-x^1)=-\delta(x^1) \tag{4.65b}$$

When $S_{12}^{\cdot\cdot 2}$ given by relation 4.64 is substituted into Eq. 4.61, we obtain

$$b^{k=2}=\int_s S_{12}^{\cdot\cdot 2}dF^{12}=\int_s S_{12}^{\cdot\cdot 2}dF^{12}+\int_s S_{21}^{\cdot\cdot 2}dF^{21} \tag{4.66}$$

However, since

$$S_{lm}^{\cdot\cdot k}=-S_{ml}^{\cdot\cdot k} \tag{4.67a}$$

and

$$dF^{12}=-dF^{21} \tag{4.67b}$$

along with the fact that

$$dF^{12}=dx^1dx^2 \tag{4.68}$$

Eq. 4.66 becomes

$$b^{k=2}=\int_{-\infty}^{+\infty}\left[\int_{-\infty}^{+\infty}\delta(x^1)\,dx^1\right]\delta(x^2)\,dx^2=1 \tag{4.69}$$

The last equation follows from the property of the Dirac delta function given by Eq. 4.5. Thus, the Burgers vector obtained from the surface integration procedure is identical to that given by the line integral technique that yields the result given by Eq. 4.38d.

Turning now to the (k') state shown in Figure 4.2, the Burgers vector can be written in terms of the following surface integral:

$$b^{k'} = -\int_S S_{l'm'}^{\cdot\cdot\ k'}\, dF^{l'm'} \tag{4.70}$$

where the torsion tensor is given by

$$S_{l'm'}^{\cdot\cdot\ k'} = A_{l'}^{L} A_{m'}^{K} \partial_{[L} A_{K]}^{k'} \tag{4.71}$$

Now there are two ways of viewing Eq. 4.70. In the first method, we treat the regions to both sides of the dislocation array separately. The only nonvanishing component of $S_{l'm'}^{\cdot\cdot\ k'}$ is given by

$$S_{12}^{\cdot\cdot 2} = \tfrac{1}{2}\bar{A}_1^1 \bar{A}_2^2 \partial_1 A_2^2 \tag{4.72}$$

which, when used along with Eq. 4.11, yields

$$S_{12}^{\cdot\cdot 2} = \left\{-\left(\tfrac{1}{2}\right)\left(\tfrac{1}{2}\right)2\delta(x^1)\right\}_1 + \left\{\left(\tfrac{1}{2}\right)\delta(x^1)\right\}_2 \tag{4.73}$$

where we have utilized

$$\bar{A}_2^2 = \left\{\tfrac{1}{2}H(-x^1)\right\}_1 + \left\{H(+x^1)\right\}_2 \tag{4.74a}$$

and

$$\bar{A}_1^1 = A_1^1 = 1 \tag{4.74b}$$

Substitution of Eq. 4.73 into Eq. 4.70 yields

$$b^{k'=2} = \int_4^3 \left[\int_{-\infty}^{+\infty} \delta(x^1)\, dx^1\right] dx^2 - \int_1^2 \left[\int_{-\infty}^{+\infty} \delta(x^1)\, dx^1\right] dx^2 \tag{4.75a}$$

which reduces to

$$b^{k'=2} = \{8\}_1 + \{-4\}_2 = 4 \tag{4.75b}$$

which is the same result as that obtained in Eq. 4.42 using the line integral method.

The second way of viewing Eq. 4.70 is in terms of a common element of area $dF^{l'm'}$, that is, say that associated with region 2; the nonplastically distorted region. Under these conditions we can write

$$A_l^L = \delta_l^L \tag{4.76}$$

in Eq. 4.71, so that Eq. 4.72 becomes

$$S_{12}^{\cdot\cdot 2} = \tfrac{1}{2}\partial_1 A_2^2 \tag{4.77}$$

or in expanded form

$$S_{12}^{\cdot\cdot 2} = \left\{-\left(\tfrac{1}{2}\right)2\delta(x^1)\right\}_1 + \left\{\left(\tfrac{1}{2}\right)\delta(x^1)\right\}_2 = \left\{\tfrac{1}{2}\delta(x^1)\right\}_1 \tag{4.78}$$

The addition of the two terms in curly brackets has become possible because of the assumption of a common element of area. Substitution of Eq. 4.78 into Eq. 4.70 yields

$$b^{k'=2} = \int_1^2 \left[\int_{-\infty}^{+\infty} \delta(x^1)\,dx^1\right] dx^2 = \{4\}_1 \tag{4.79}$$

that again is the same result as that given by Eq. 4.75b. The common area could also have been taken with respect to region 2, that is, the left side or plastically distorted region of Figure 4.2.; however, $S_{12}^{\cdot\cdot 2}$ would need to be written in the form given by Eq. 4.72 so that Eq. 4.78 would become

$$S_{12}^{\cdot\cdot 2} = \left\{-\left(\tfrac{1}{2}\right)\left(\tfrac{1}{2}\right)2\delta(x^1)\right\}_1 + \left\{\left(\tfrac{1}{2}\right)\left(\tfrac{1}{2}\right)\delta(x^1)\right\}_2 = \left\{-\left(\tfrac{1}{2}\right)\left(\tfrac{1}{2}\right)\delta(x^1)\right\}_1 \tag{4.80}$$

The extra factor of (1/2) in the above equation arises from the term $\bar{A}_2^2$ in Eq. 4.72. However, $S_{12}^{\cdot\cdot 2}$ must now be integrated in the dislocated region, that is, over the distance

$$\int_4^3 dx^2 = 8 \tag{4.81}$$

so that Eq. 4.80, when substituted into Eq. 4.70, yields the same result as Eq. 4.79. The representation of the torsion tensor in terms of Eq. 4.78 has a somewhat more fundamental significance and is therefore used throughout the remaining equations.

The following surface integral yields the Burgers vector associated with the (k^1) state of Figure 4.3*b*

$$b^{k^1} = -\int_s S_{l^1 m^1}^{\cdot\cdot\, k^1}\, dF^{l^1 m^1} \tag{4.82}$$

where

$$S_{l^1 m^1}^{\cdot\cdot\, k^1} = A_{l^1}^{L} A_{m^1}^{K} \partial_{[L} A_{K]}^{k^1} \tag{4.83}$$

Using Eqs. 4.14, the only nonvanishing component of $S_{l^1 m^1}^{\cdot\cdot\ k^1}$ becomes

$$S_{12}^{\cdot\cdot 2} = -\left(\tfrac{1}{2}\right)\left[-\delta(x^2)+1\right]\delta(x^1)+\tfrac{1}{2}\delta(x^1)=\tfrac{1}{2}\delta(x^1)\delta(x^2) \tag{4.84}$$

which when substituted into Eq. 4.82, yields

$$b^{k^1=2} = -1 \tag{4.85}$$

which again is the same result as that given by Eq. 4.45b using the line integral method.

For the $(k^{1'})$ state of Figure 4.3*c*, we can write for the closure failure

$$b^{k^{1'}} = -\int_s S_{l^{1'}m^{1'}}^{\cdot\cdot\ k^{1'}}\, dF^{l^{1'}m^{1'}} \tag{4.86}$$

where

$$S_{l^{1'}m^{1'}}^{\cdot\cdot\ k^{1'}} = A_{l^{1'}}^{L} A_{m^{1'}}^{K} \partial_{[L} A_{K]}^{k^{1'}} \tag{4.87}$$

and where the only nonvanishing component of $S_{l^{1'}m^{1'}}^{\cdot\cdot\ k^{1'}}$ is given by

$$S_{12}^{\cdot\cdot 2} = \left\{-\left(\tfrac{1}{2}\right)\left(\tfrac{1}{2}\right)\delta(x^1)\right\}_1 + \left\{\left(\tfrac{1}{2}\right)\delta(x^1)\right\}_2 = \left\{\left(\tfrac{1}{4}\right)\delta(x^1)\right\}_2 \tag{4.88}$$

where the distortions given by Eq. 4.15 have been used. Substituting Eq. 4.88 into Eq. 4.86 yields

$$b^{k^{1'}=2} = \{-2\}_2 \tag{4.89}$$

which is the same result obtained for Eq. 4.47 using the line integral procedure.

Turning to the (k^2) state of Figure 4.4*b*, the closure failure may be written as

$$b^{k^2} = -\int_s S_{l^2 m^2}^{\cdot\cdot\ k^2}\, dF^{l^2 m^2} \tag{4.90}$$

where

$$S_{l^2 m^2}^{\cdot\cdot\ k^2} = A_{l^2}^{L} A_{m^2}^{K} \partial_{[L} A_{K]}^{k^2} \tag{4.91}$$

and where the only nonvanishing component of $S_{l^2 m^2}^{\cdot\cdot\ k^2}$ can be written as

$$S_{12}^{\cdot\cdot 1} = -\tfrac{1}{2}\bar{A}_1^1 \bar{A}_2^2 \partial_2 A_1^1 \tag{4.92}$$

which in view of Eqs. 4.17 becomes

$$S_{12}^{\cdot\cdot 1} = -\left(\tfrac{1}{2}\right)\left[\delta(x^1)+1\right]\delta(x^2)+\tfrac{1}{2}\delta(x^2) = -\tfrac{1}{2}\delta(x^1)\delta(x^2) \tag{4.93}$$

and which when substituted into Eq. 4.90 yields

$$b^{k^2=1} = +1 \tag{4.94}$$

This is precisely the value given by Eq. 4.49c using the line integral technique.

In the case of the closure failure associated with the $(k^{2'})$ state shown in Figure 4.5, we may write

$$b^{k^{2'}} = -\int_s S_{l^{2'}m^{2'}}^{\cdot\cdot\; k^{2'}} dF^{l^{2'}m^{2'}} \tag{4.95}$$

where the torsion tensor is given by

$$S_{l^{2'}m^{2'}}^{\cdot\cdot\; k^{2'}} = A_{l^{2'}}^{L} A_{m^{2'}}^{K} \partial_{[L} A_{K]}^{k^{2'}} \tag{4.96}$$

The only nonvanishing component of $S_{l^{2'}m^{2'}}^{\cdot\cdot\; k^{2'}}$ is found to be

$$S_{12}^{\cdot\cdot 1} = \tfrac{1}{2}\bar{A}_1^1 \bar{A}_2^2 \partial_1 A_2^1 \tag{4.97}$$

which in view of Eq. 4.18 yields

$$S_{12}^{\cdot\cdot 1} = -\left\{\left(\tfrac{1}{2}\right)\delta(x^1)\tan\theta\right\}_1 \tag{4.98}$$

When the above equation is substituted into Eq. 4.95, we obtain

$$b^{k^{2'}=1} = \int_4^3 \left[\int_{-\infty}^{+\infty} \delta(x^1)\,dx^1\right] \tan\theta\, dx^2 = 4\tan\theta = 2 \tag{4.99}$$

which is the same result as that given by the line integral method that led to Eq. 4.51c.

Finally, we can write for the (k'^c) state shown in Figure 4.6

$$b^{k'^c} = -\int_s S_{l'^c m'^c}^{\cdot\cdot\; k'^c} dF^{l'^c m'^c} \tag{4.100}$$

where

$$S_{l'^c m'^c}^{\cdot\cdot\; k'^c} = A_{l'^c}^{L} A_{m'^c}^{K} \partial_{[L} A_{K]}^{k'^c} \tag{4.101}$$

There are now two nonvanishing components of $S_{l'^c m'^c}^{\cdot\;\cdot\;k'^c}$ that can be written as

$$S_{12}^{\cdot\;\cdot\;1}=\tfrac{1}{2}\partial_1 A_2^1\left[\bar{A}_1^1\bar{A}_2^2-\bar{A}_1^2\bar{A}_2^1\right] \tag{4.102a}$$

and

$$S_{12}^{\cdot\;\cdot\;2}=\tfrac{1}{2}\partial_1 A_2^2\left[\bar{A}_1^1\bar{A}_2^2-\bar{A}_1^2\bar{A}_2^1\right] \tag{4.102b}$$

which in view of Eq. 4.35 yields

$$\begin{aligned} S_{12}^{\cdot\;\cdot\;1}&=\left\{-\left(\tfrac{1}{2}\right)\sqrt{2}\,\sin\theta\delta(x^1)\right\}_1+\left\{\left(\tfrac{1}{2}\right)\left(\tfrac{1}{2}\right)\sqrt{2}\,\sin\theta\delta(x^1)\right\}_2\\ &=\left\{-\left(\tfrac{1}{4}\right)\sqrt{2}\,\sin\theta\delta(x^1)\right\}_1 \end{aligned} \tag{4.103a}$$

$$\begin{aligned} S_{12}^{\cdot\;\cdot\;2}&=\left\{-\left(\tfrac{1}{2}\right)\sqrt{2}\,\cos\theta\,\delta(x^1)\right\}_1+\left\{\left(\tfrac{1}{2}\right)\left(\tfrac{1}{2}\right)\sqrt{2}\,\cos\theta\,\delta(x^1)\right\}_2\\ &=\left\{-\left(\tfrac{1}{4}\right)\sqrt{2}\,\cos\theta\,\delta(x^1)\right\}_1 \end{aligned} \tag{4.103b}$$

An alternate way of obtaining the above results is to transform the components $S_{l'm'}^{\cdot\;\cdot\;k'}$ in the (k') system as given by Eq. 4.78 into the components $S_{l'^c m'^c}^{\cdot\;\cdot\;k'^c}$ in the new system (k'^c). This can be done by means of the following coordinate transformation:

$$S_{l'^c m'^c}^{\cdot\;\cdot\;k'^c}=C_{l'^c}^{l'}C_{m'^c}^{m'}C_{k'}^{k'^c}S_{l'm'}^{\cdot\;\cdot\;k'} \tag{4.104}$$

where

$$C_{k'}^{k'^c}=\delta_{K^c}^{k'^c}\,\delta_{k'}^{K}C_K^{K^c} \tag{4.105}$$

where $C_K^{K^c}$ is given by Eq. 4.20. Upon expanding Eq. 4.104, we obtain

$$S_{12}^{\cdot\;\cdot\;1}=\bar{C}_1^1\bar{C}_2^2C_2^1S_{12}^{\cdot\;\cdot\;2}-\bar{C}_1^2\bar{C}_2^1C_2^1S_{12}^{\cdot\;\cdot\;2} \tag{4.106a}$$

and

$$S_{12}^{\cdot\;\cdot\;2}=\bar{C}_1^1\bar{C}_2^2C_2^2S_{12}^{\cdot\;\cdot\;1}-\bar{C}_1^2\bar{C}_2^1C_2^2S_{12}^{\cdot\;\cdot\;2} \tag{4.106b}$$

Since $S_{12}^{\cdot\;\cdot\;2}$ in the last two expressions is given by Eq. 4.78, it is clear that Eq. 4.106 are identical to those given by Eq. 4.103. We can now substitute Eq. 4.103 into Eq. 4.100 to obtain $b^{k'^c}$; however, before doing so we must convert the Dirac delta function $\delta(x^K)$ into that associated with the new coordinate system (K^c). This is done by first writing

$$\frac{\partial H(x^L)}{\partial x^K}=\frac{\partial H\left(A_{K^c}^{L}x^{K^c}\right)}{\partial x^{L^c}}\frac{\partial x^{L^c}}{\partial x^K} \tag{4.107}$$

The nonvanishing component of the above equation is

$$\delta(x^1) = \frac{\partial H(x^1)}{\partial x^1} = \frac{\partial H(A^1_{K^c} x^{K^c})}{\partial x^{L^c}} \frac{\partial x^{L^c}}{\partial x^1} \tag{4.108a}$$

that can be expanded as

$$\delta(x^1) = \frac{\partial H(\overline{A}^1_1 x^1 + \overline{A}^1_2 x^2)}{\partial x^1} \frac{\partial x^1}{\partial x^1} + \frac{\partial H(\overline{A}^1_1 dx^1 + \overline{A}^1_2 x^2)}{\partial x^2} \frac{\partial x^2}{\partial x^1} \tag{4.108b}$$

or still further since $\partial x^1/\partial x^1$ and $\partial x^2/\partial x^1$ correspond to A^1_1 and A^2_1, respectively,

$$\delta(x^1) = A^1_1 \delta(x^1) + A^2_1 \delta(x^2) \tag{4.109}$$

where $\overline{A}^1_1$ and $\overline{A}^1_2$ have been omitted since they do not affect the Heaviside function because they appear only as factors. We can now utilize Eq. 4.109, together with Eqs. 4.103 and 4.100 to obtain

$$b^{k'^c=1} = \left\{ 2\left(\tfrac{1}{4}\right)\sqrt{2}\, \sin\theta \left[\int_1^2 \delta(x^1)\, dx^1 + \int_1^2 \delta(x^2)\, dx^2 \right] \right\}_1 \tag{4.110}$$

and

$$b^{k'^c=2} = \left\{ 2\left(\tfrac{1}{4}\right)\sqrt{2}\, \cos\theta \left[\int_1^2 \delta(x^1)\, dx^1 + \int_1^2 \delta(x^2)\, dx^2 \right] \right\}_1 \tag{4.111}$$

Since θ here is chosen as 45°, it is clear that the results given by Eq. 4.110 are identical to those given by Eq. 4.55 that were obtained by the line integral method.

It is also clear at this point that the Burgers vector associated with state (k'^c) can be obtained from that in state (k') by the following coordinate transformation:

$$b^{k'^c} = C^{k'^c}_{k'} b^{k'} \tag{4.112}$$

so that

$$b^{k'^c=1} = C^1_2 b^{k'=2} \tag{4.113}$$

and

$$b^{k'^c=2} = C^2_2 b^{k'=2} \tag{4.114}$$

where $b^{k'=2}$ is given by Eq. 4.42. It follows then that Eqs. 4.113 and 4.114 are identical to those given by Eq. 4.110.

We are now in a position to note that whenever a given state has associated with it a nonvanishing torsion tensor, then that state contains dislocations. This very important result was first proposed by Kondo (1955).

The surface integral given by Eq. 4.61 can now be written in differential form to yield (Kröner, 1958)

$$db^k = -S_{lm}^{\cdot\cdot k}\, dF^{lm} = -S_{lm}^{\cdot\cdot k}\varepsilon^{nlm}\, dF_n = \alpha^{nk}\, dF_n \tag{4.115}$$

In the above relation, use has been made of the fact that

$$dF^{lm} = \varepsilon^{nlm}\, dF_n \tag{4.116}$$

where ε^{nlm} is termed the permutation tensor. In particular, the permutation tensor always allows one to associate a given vector dF_n with an antimetric tensor (Aris, 1962) dF^{lm}. The tensor dF^{lm} is antimetric since

$$dF^{lm} = -dF^{ml} \tag{4.117}$$

The quantity α^{nk} in Eq. 4.115 is termed the dislocation density tensor and may be written as

$$\alpha^{nk} = -\varepsilon^{nlm} S_{lm}^{\cdot\cdot k} \tag{4.118}$$

The permutation symbol ε^{nlm} can further be written as

$$\varepsilon^{nlm} = e^{nlm}/\sqrt{a} \tag{4.119}$$

where e^{nlm} is the permutation symbol defined such that it is equal to $+1$ or -1, depending on whether $n, l, m = 1, 2, 3$ is an even or odd permutation respectively of this sequence and 0 if any two or three of the superscripts are equal. The quantity a is the determinant of the metric tensor associated with the given state.

For the specific case of the (k) state shown in Figure 4.1b, Eq. 4.118 yields

$$\alpha^{32} = -2S_{12}^{\cdot\cdot 2} = \delta(x^2)\,\delta(x^1) \tag{4.120}$$

where the torsion tensor utilized in the last expression is that given by Eq. 4.64, while $a = 1$. In words, Eq. 4.120 states that the dislocation density is everywhere zero except right at the dislocation core. Upon integration, Eq.

4.120 yields the value of $+1$, which in words means that there is one extra half plane at the position of the dislocation.

For the (k') state shown in Figure 4.2, we can write

$$\alpha^{n'k'} = -\varepsilon^{n'l'm'} S_{l'm'}^{\cdot\;\cdot\;k'} \tag{4.121}$$

Now as we have seen earlier, there are three ways of writing $S_{l'm'}^{\cdot\;\cdot\;k'}$. Using Eq. 4.73, we obtain

$$\alpha^{32} = -2S_{12}^{\cdot\;\cdot\;2}/\sqrt{a} = \{2\delta(x^1)\}_1 + \{-\delta(x^1)\}_2 = \{\delta(x^1)\}_1 \tag{4.122}$$

where a is the determinant associated with the following metric tensor

$$a_{k'l'} = A_{k'}^{K} A_{l'}^{L} \delta_{KL} \tag{4.123}$$

In view of Eq. 4.74, a associated with the leftmost curly bracket in Eq. 4.122 is $1/4$, while for the rightmost curly bracket, it is equal to 1. Again, it should be emphasized that $a_{k'l'}$ is not the metric tensor characteristic of the (k') state, since this metric is given by Eq. 4.12. When Eq. 4.78 is used for the torsion tensor, Eq. 4.121 yields

$$\alpha^{32} = -S_{12}^{\cdot\;\cdot\;2}/\sqrt{a} = \{\delta(x^1)\}_1 \tag{4.124}$$

where a is equal to 1. When $S_{12}^{\cdot\;\cdot\;2}$ given by Eq. 4.80 is substituted into Eq. 4.121, we obtain the same result as given by Eq. 4.124, since a in this case equals $1/4$. Again for convenience, we shall express the dislocation density in the form given by Eq. 4.124. Thus, the dislocation density vanishes everywhere except at $x^1 = 0$, that is, the origin of the vertical dislocation array in Figure 4.2. Upon integration, Eq. 4.124 yields $\alpha^{32} = 1$, which merely states that there is one extra half plane for each plane of the original undistorted reference lattice. Note also that the first superscript in α^{32} corresponds to the direction of the dislocation line, that is, normal to the plane of the drawing, while the second superscript denotes the direction of the Burgers vector, that is, the vertical direction in Figure 4.2.

Returning again to the (k^1) state shown in Figure 4.3*b*, we can write

$$\alpha^{n^1k^1} = -\varepsilon^{n^1l^1m^1} S_{l^1m^1}^{\cdot\;\cdot\;k^1} \tag{4.125}$$

which in view of Eq. 4.84 can be written as

$$\alpha^{32} = -2S_{12}^{\cdot\;\cdot\;2} = -\delta(x^2)\delta(x^1) \tag{4.126}$$

which when integrated yields -1. Physically, this result means that there is one missing plane per plane of the original undistorted crystal.

In the case of the $(k^{1'})$ state of Figure 4.3*c* we can write

$$\alpha^{n^{1'}k^{1'}} = -\varepsilon^{n^{1'}l^{1'}m^{1'}} S_{l^{1'}m^{1'}}^{\cdot\cdot\ k^{1'}} \tag{4.127}$$

where the only nonvanishing component yields in view of Eq. 4.88

$$\alpha^{32} = -2S_{12}^{\cdot\cdot 2}/a = \left\{-\left(\tfrac{1}{2}\right)\delta(x^1)\right\}_2 \tag{4.128}$$

In words, the above equation when integrated says that there is one missing plane for every two planes existing in the initial reference state crystal.

For the (k^2) state of Figure 4.4*b*, we have

$$\alpha^{n^2k^2} = -\varepsilon^{n^2l^2m^2} S_{l^2m^2}^{\cdot\cdot\ k^2} \tag{4.129}$$

or in view of Eq. 4.93

$$\alpha^{31} = -2S_{12}^{\cdot\cdot 1} = \delta(x^1)\,\delta(x^2) \tag{4.130}$$

which again states that there is a single extra half plane per plane of the original reference crystal. Similarly for the $(k^{2'})$ state of Figure 4.5, the dislocation density tensor becomes

$$\alpha^{n^{2'}k^{2'}} = -\varepsilon^{n^{2'}l^{2'}m^{2'}} S_{l^{2'}m^{2'}}^{\cdot\cdot\ k^{2'}} \tag{4.131}$$

or in view of Eq. 4.98

$$\alpha^{31} = -2S_{12}^{\cdot\cdot 1}/a = \{\delta(x^1)\tan\theta\}_1 \tag{4.132}$$

where a is found to be unity. In terms of Figure 4.5 the integrated form of Eq. 4.132 simply gives the number of half planes, that is, 2, for every four horizontal planes along the line 4–3 associated with the original undeformed lattice.

Finally, for the (k'^c) state of Figure 4.6, we have

$$\alpha^{n'^c k'^c} = -\varepsilon^{n'^c l'^c m'^c} S_{l'^c m'^c}^{\cdot\cdot\ k'^c} \tag{4.133}$$

Using the values of $S_{l'^c m'^c}^{\cdot\cdot\ k'^c}$ given by Eq. 4.103, we find

$$\alpha^{31} = -2S_{12}^{\cdot\cdot 1} = \left\{\left(\tfrac{1}{2}\right)\delta(x^1)\right\}_2 \tag{4.134a}$$

and

$$\alpha^{32} = -2S_{12}^{\cdot\cdot 2} = \left\{\left(\tfrac{1}{2}\right)\delta(x^1)\right\}_1 \tag{4.134b}$$

or upon integration after using Eq. 4.109

$$\alpha^{31} = \alpha^{32} = \{1\}_1 \tag{4.135}$$

These dislocation densitities correspond to the presence of one extra half plane per full plane of the (k'^c) state lattice along both the x^1 and x^2 directions. The dislocation density tensor $\alpha^{n'^c k'^c}$ could alternately be obtained using the following transformation:

$$\alpha^{n'^c k'^c} = C_{n'}^{n'^c} C_{k'}^{k'^c} \alpha^{n'k'} \tag{4.136}$$

where $\alpha^{n'k'}$ and $C_{k'}^{k'^c}$ are given by Eqs. 4.122 and 4.105, respectively.

REVIEW

The dislocations considered in this chapter all terminate within the interior of the crystal. In a sense, we may therefore say that the plastic distortion terminates within the body, that is, is imperfect. The distortion tensors that described the dislocation content are thus discontinuous. In particular, for the planar dislocation array shown in Figure 4.2, the distortion tensor is in the form of a Heaviside function $H(x^1)$ given by Eq. 4.11 where these functions are defined by Eq. 4.3. It follows from these equations that the Heaviside functions are step functions that confine the plastic distortion to only one area of the body. On the other hand, for a single dislocation, such as depicted in Figure 4.1*b*, the distortions take the form $H(x^1)\delta(x^2)$ where the $\delta(x^2)$ are Dirac delta functions defined by Eqs. 4.4 and 4.5. They serve to concentrate the dislocation at one point in the body, that is, to quantize it.

Consider a closed reference circuit made in the perfect reference crystal (K) and the distortion tensor A_K^k, giving the dislocated state (k) integrated about such a circuit. Then a closure failure b^k giving the total dislocation content contained therein can be obtained. A typical integral of this type is given by Eq. 4.37. The distortion tensor associated with the dislocated (k) states can be used to obtain yet another quantity $S_{lm}^{\cdot\cdot k}$ termed the torsion tensor as defined by Eq. 4.62. The torsion tensor is seen to be a function of $\partial_{[L} A_{K]}^k$ where the brackets designate the asymmetric components of L and K. Thus, for a continuous function A_K^k these derivatives vanish, whereas for a discontinuous function, such as occurs for dislocations, they possess finite values. In particular, when the Heaviside step functions are differentiated, they become Dirac delta functions in accordance with Eq. 4.65. Thus, the nonvanishing of the torsion tensor is directly related to the discrete or discontinuous nature of a dislocation. It may also be viewed as the most important tensor quantity associated with a dislocation in that its presence is synonymous with the presence of dislocations, whereas its vanishing corresponds to the absence of dislocations. The vanishing of $S_{lm}^{\cdot\cdot k}$ also defines a Riemannian or nondislocated space, whereas its presence corresponds to a more general and therefore more complex non-Riemannian space. It can also be pointed out here that Einstein's general theory of relativity and nearly all subsequent developments of this theory have concerned themselves with the more restrictive Riemannian geometries. It is therefore to be anticipated that a number of exciting discoveries will be made upon application of the more general non-Riemannian geometry to the four-dimensional space-time continuum.

Analogous to the way in which the distortion tensor can be integrated about a closed path to yield the total dislocation content within this path, Stokes' theorem can be used to obtain a corresponding surface integral counterpart with which to determine the Burgers vector. In this case, the quantity to be integrated is the torsion tensor. A typical example is given by Eq. 4.61. If, on the other hand, this surface integral is differentiated with respect to the area, we obtain the dislocation density tensor α^{nk}, which, as Eq. 4.118 shows, is just the torsion tensor. The types of dislocation configurations needed to generate most of the plastic states discussed in the previous chapter have been treated in detail with respect to their various related tensor quantities. The following chapter will describe in detail what happens to an internal dislocation when it reaches the surface of a crystal.

CHAPTER 5

Free Surface Boundary Conditions

Nearly all of the developments in dislocation theory have been made with respect to infinite bodies. This of course simplifies the calculations, but at the same time it has postponed a real understanding of the nature of a free surface, that is, the boundary value problem associated with bodies subjected either to internal or external stresses. In this chapter we show for the first time that the stress-free boundary conditions can be satisfied for any state of stress by the use of some suitable distribution of surface dislocations. This is an extremely powerful technique that not only yields numerical results, but also leads to considerable physical insight with respect to a given problem.

5.1 DISLOCATIONS AND FREE SURFACES

It is clear from Figure 4.2 that when the dislocations associated with the (k') state move, or more accurately, climb toward the rightmost surface of the crystal, the fully plastic (a) state shown in Figure 3.1a obtains. Under these conditions all of the elastic strain is removed. In view of the Burgers circuit shown in Figure 4.2, it seems logical to construct the Burgers vector for the (a) state in the manner shown in Figure 5.1. In particular the four closure failures along the line 4–3 still remain and are of course associated with the four dislocations. In addition, however, we now have four closure failures along the line 1–2. These may be viewed as being created by the formation of new surface by the four dislocations (Marcinkowski, 1978b, 1978c, 1978d). These surface closure failures are seen to have opposite signs to those of the dislocations. It thus seems reasonable to write the closure failures for the two contributions as

$$b^a = -\oint A_K^a \, dx^K \tag{5.1a}$$

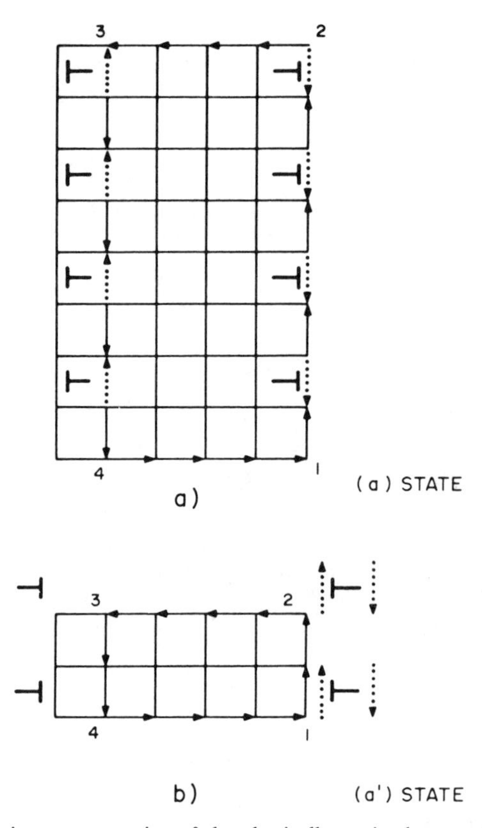

Figure 5.1 Dislocation representation of the plastically strained states shown in Figure 3.1.

and

$$\underset{s}{b}^{a} = \oint A_K^a \, dx^K \tag{5.1b}$$

where b^a corresponds to the closure failure associated with the dislocations, while $\underset{s}{b}^{a}$ corresponds to the closure failure associated with the additional free surfaces created by these dislocations. The distortion in Eq. 5.1 is given by

$$A_K^a = \delta_{k'}^a \delta_K^L A_L^{k'} \tag{5.2}$$

where $A_L^{k'}$ is given by Eq. 4.11. It is now important to note that x^1 in the corresponding Heaviside functions must now be measured with respect to the rightmost face of the crystal in Figure 1a. Strictly speaking, however, the line 1–2 in Figure 5.1a lies towards the right of $x^1=0$, but infinitesimally close to it. Combining Eqs. 5.1 and 5.2 yields

$$b^{a=2} = -\int_1^2 dx^2 - 2\int_3^4 dx^2 = (-4) + (8) = 4 \tag{5.3a}$$

and

$$\underset{s}{b}^{a=2} = \int_1^2 dx^2 + 2\int_3^4 dx^2 = (4) + (-8) = -4 \tag{5.3b}$$

These two equations give the closure failures along the lines 4–3 and 2–1, respectively, in Figure 5.1a. The surface closure failure given by Eq. 5.1b could alternately have been written as

$$\underset{s}{b}^{a} = -\oint \underset{s}{A}{}_K^a \, dx^K \tag{5.4}$$

where now the component A_2^2 in Eq. 4.11b would have to be written as

$$\underset{s}{A}{}_2^2 = \{H(-x^1)\}_1 + \{2H(+x^1)\}_2 \tag{5.5}$$

When we consider the (a^1) state as shown in Figure 3.1b and try to construct a Burgers circuit in such a crystal, the configuration shown in Figure 5.1b obtains. This construction is apparent when we consider the $(k^{1'})$ state shown in Figure 4.3c. In particular, the two downward pointing dotted arrows towards the rightmost face of the crystal in Figure 5.1b correspond to the two dislocations that have left the crystal. On the other hand, the two upward pointing arrows correspond to the two free surfaces associated with these dislocations that have been removed from the original crystal. We can thus write

$$b^{a^1} = -\oint A_K^{a^1} \, dx^K \tag{5.6a}$$

and

$$\underset{s}{b}^{a^1} = \oint A_K^{a^1} \, dx^K \tag{5.6b}$$

where

$$A_K^{a^1} = \delta_{k^{1'}}^{a^1} \delta_K^L A_L^{k^{1'}} \tag{5.7}$$

and where $A_L^{k^{1'}}$ is given by Eq. 4.15.

Combining Eqs. 5.6 and 5.7, we obtain

$$b^{a^1=2} = -\int_1^2 dx^2 - \frac{1}{2}\int_3^4 dx^2 = (-4) + (2) = -2 \tag{5.8a}$$

and

$$\underset{s}{b}^{a^1=2} = \int_1^2 dx^2 + \frac{1}{2}\int_3^4 dx^2 = (4) + (-2) = 2 \tag{5.8b}$$

These two results correspond to closure failures shown in Figure 5.1*b*.

The dislocation representation of the plastically strained state shown in Figure 3.2*a* is illustrated in Figure 5.2*a*. The more continuous dislocation representation of this state is shown in Figure 5.2*b* and corresponds to the (a^2) state of Figure 3.2*b*. We note that the Burgers circuit 1–1′–2–2′–3–4–1 consists of the closure failure 2′–2 due to dislocations, while that given by 1′–1 corresponds to the free surface contribution associated with these dislocations. Quantitatively, we may write

$$b^{a^2} = -\oint A_K^{a^2} dx^K \tag{5.9a}$$

and

$$\underset{s}{b}^{a^2} = \oint A_K^{a^2} dx^K \tag{5.9b}$$

where

$$A_K^{a^2} = \delta_{k^{2'}}^{a^2} \delta_K^L A_L^{k^{2'}} \tag{5.10}$$

and where $A_L^{k^{2'}}$ is given by Eq. 4.18. Combining Eqs. 5.9 and 5.10 we obtain

$$b^{a^2=1} = -\int_3^4 A_2^1 dx^2 = 4\tan\theta = 2 \equiv \underset{2'-2}{\Delta x^1} \tag{5.11a}$$

and

$$\underset{s}{b}^{a^2=1} = \int_3^4 A_2^1 dx^2 = -4\tan\theta = -2 \equiv \underset{1'-1}{\Delta x^1} \tag{5.11b}$$

which is in agreement with the closure failures shown in Figure 5.2*b*.

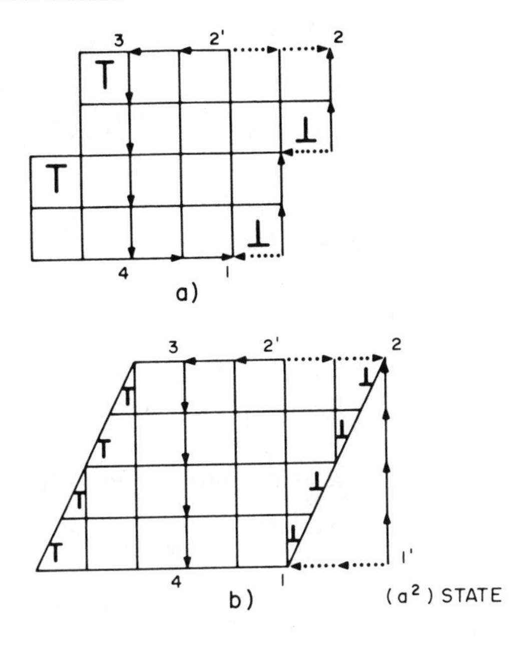

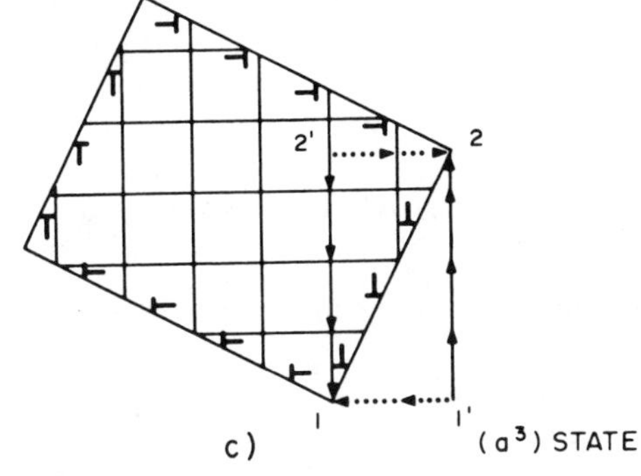

Figure 5.2 Dislocation representation of the plastically strained states shown in Figure 3.2.

Along the lines of Eqs. 5.4 and 5.5, an alternate way of writing Eq. 5.9b is

$$\underset{s}{b}^{a^2} = -\oint \underset{s}{A}_K^{a^2}\, dx^K \tag{5.12}$$

where

$$\underset{s}{A}_2^1 = \{\tan\theta H(+x^1)\}_2 \tag{5.13}$$

The dislocation representation of the pure plastic rotation of Figure 3.2*c* is shown in Figure 5.2*c*. In particular, the 1–2 face of the drawing is seen to contain dislocations of total strength 2′–2 that possesses a corresponding free surface or ledge strength given by 1′–1. These contributions are given quantitatively by

$$b^{a^3} = -\oint A_K^{a^3}\, dx^K \tag{5.14a}$$

and

$$\underset{s}{b}^{a^3} = \oint \underset{s}{A}_K^{a^3}\, dx^K \tag{5.14b}$$

where the distortions used in the above equations must be written as

$$\underset{s}{A}_K^{a^3} = \left\{ A_K^{a^3} H(-x^1) \right\}_{,1} \tag{5.15}$$

where $A_K^{a^3}$ within curly brackets is given by Eq. 3.18. Combining Eq. 5.15 with Eq. 5.14 yields

$$b^{a^3=1} = -\int_3^4 A_2^1\, dx^2 = 4\sin\theta \equiv \underset{2'-2}{\Delta x^1} \tag{5.16a}$$

and

$$\underset{s}{b}^{a^3=1} = \int_3^4 \underset{s}{A}_2^1\, dx^2 = -4\sin\theta \equiv \underset{1'-1}{\Delta x^1} \tag{5.16b}$$

which is in agreement with the results of Figure 5.2*c*. Note that although we could write

$$b^{a^3=2} = -\int_3^4 A_2^2\, dx^2 = 4\cos\theta \equiv \underset{1-2'}{\Delta x^2} \tag{5.17a}$$

and

$$\underset{s}{b}^{a^3=2} = \int_3^4 \underset{s}{A}_2^2\, dx^2 = -4\cos\theta \equiv \underset{1'-2}{\Delta x^2} \tag{5.17b}$$

these do not correspond to closure failures associated with newly created free surfaces or new extra half planes, but rather correspond to those planes and surfaces existing in the original reference crystal. Thus, Eq. 5.17 does not represent valid closure failures and are thus indicated by solid arrows in Figure 5.2*c*. Suitable Burgers circuits could also be made about

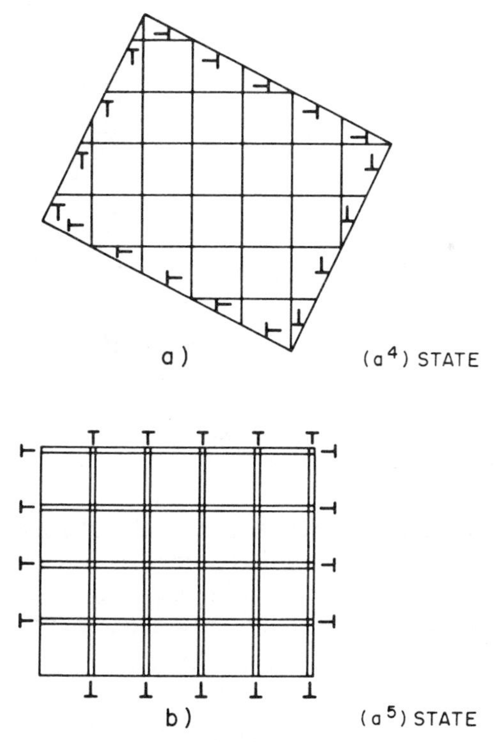

Figure 5.3 Dislocation representation of the plastically strained states shown in Figure 3.3.

the three remaining faces of the (a^3) state crystal and the dislocation content readily obtained.

Figures 5.3*a*, 5.3*b*, and 5.4 show the dislocation representations of the fully plastic states corresponding to those shown in Figures 3.3*a*, 3.3*b*, and 3.4, respectively. In particular, the closure failure analysis for the (a^4) and (a^6) states is similar to that carried out for the (a^2) state, while that for the (a^5) state is similar to that made for the (a) state.

Interesting modifications of the (a) and (a^2) states shown in Figures 5.1*a* and 5.2*b*, respectively, are given in Figures 5.5*a* and 5.5*b*, respectively. In these latter cases, all of the extra surfaces at the rightmost face of the crystal associated with the dislocations at this surface have been omitted. As is shown in a later section, such behavior could occur if the free surface energy associated with the formation of these surfaces approached infinity. It is shown that internal elastic strains are associated with these configurations. It follows that the surface dislocation failures $b^{a=2}$ and $b^{a^2=1}$, shown dotted in Figures 5.5*a* and 5.5*b*, remain the same as those given by Eqs.

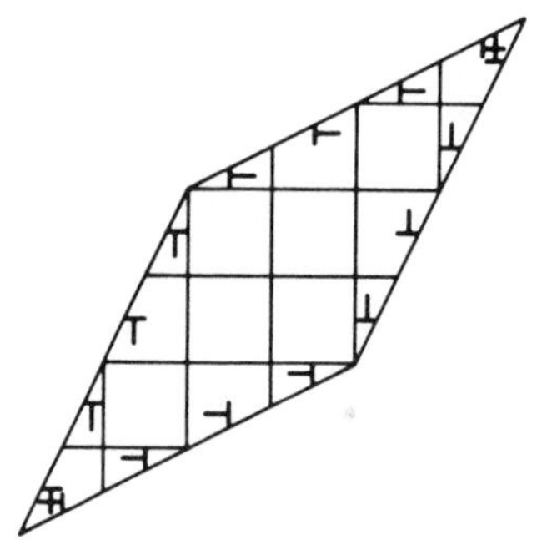

Figure 5.4 Dislocation representation of the plastically strained state shown in Figure 3.4.

5.3a and 5.11a, respectively. On the other hand, $\underset{s}{b}^{a=2}$ and $\underset{s}{b}^{a^2=1}$ now become equal to zero.

Finally, the dislocation representation of the plastically strained state given in Figure 4.7*b* is shown in Figure 5.6. This is obviously the same state as that shown in Figure 5.1*a*, but expressed in terms of a different coordinate system. The closure failures associated with this particular state could be written as

$$b^{a^c} = -\oint A_K^{a^c}\, dx^K \tag{5.18a}$$

and

$$\underset{s}{b}^{a^c} = \oint A_K^{a^c}\, dx^K \tag{5.18b}$$

where

$$A_K^{a^c} = \delta_{k'^c}^{a^c} \delta_K^L A_L^{k'^c} \tag{5.19}$$

and where $A_L^{k'^c}$ is given by Eq. 4.35 and where it must be remembered that x^1 in the Heaviside functions are now measured with respect to the rightmost face of Figure 5.6.

It is instructive at this point to view the dislocation contents associated with Figure 5.5 from a somewhat different point of view. In particular, the Burgers circuits shown in Figures 5.5*a* and 5.5*b* can be redrawn as shown in Figures 5.7*a* and 5.7*b*, respectively. For convenience, those circuits may be viewed as shrunk in size so that they may be expressed in differential form. The closure failures, that are shown dotted in Figure 5.7 may be

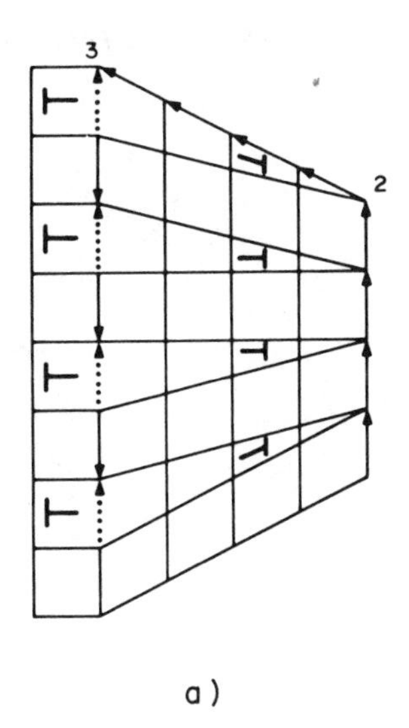

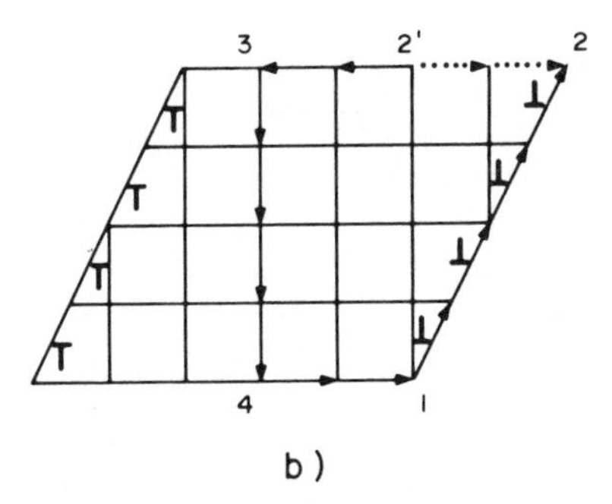

Figure 5.5 The distortions shown in *a*) and *b*) are the same as those shown in Figures 5.1*a* and 5.2*b*, respectively, except that now there are no extra free surfaces associated with the rightmost face 1–2.

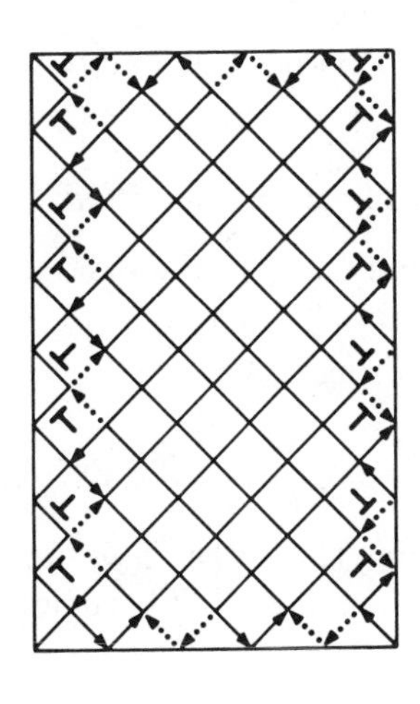

Figure 5.6 Dislocation representation of the plastically strained state shown in Figure 4.7*b*.

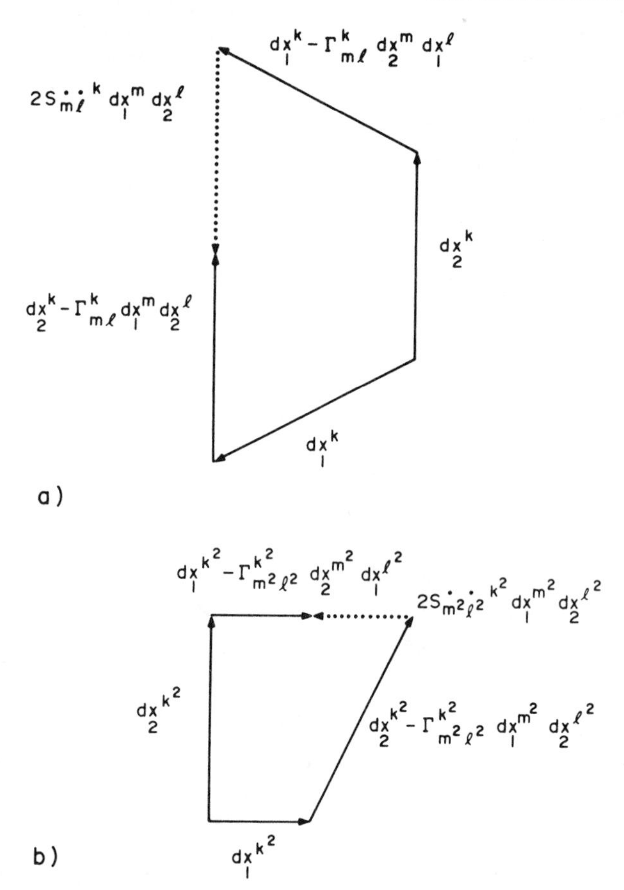

Figure 5.7 Interpretation *a*) of Figure 5.5*a* and *b*) of Figure 5.5*b* in terms of the concept of parallel displacement.

viewed as generated by the displacement of a pair of vectors, that is, $\underset{1}{dx}^k$ and $\underset{2}{dx}^k$ in the case of Figure 5.7*a* and $\underset{1}{dx}^{k^2}$ and $\underset{2}{dx}^{k^2}$ in the case of Figure 5.7*b*. In the most general case, the differential of x^k must be written as a covariant differential, that is (Schouten, 1954; Eringen, 1966),

$$\delta x^k \overset{\text{def}}{=} dx^k + \Gamma^k_{ml} x^l dx^m \tag{5.20}$$

where Γ^k_{ml} are termed the coefficients of connection. Thus, when a vector $\underset{2}{dx}^l$ is displaced in a parallel manner through a distance $\underset{1}{dx}^m$, it becomes

$$\underset{2}{dx}^k - \Gamma^k_{ml}\, \underset{1}{dx}^m\, \underset{2}{dx}^l \tag{5.21a}$$

while a vector $\underset{1}{dx}^l$, when displaced in a parallel manner through $\underset{2}{dx}^m$ becomes

$$\underset{1}{dx}^k - \Gamma^k_{ml}\, \underset{2}{dx}^m\, \underset{1}{dx}^l \tag{5.21b}$$

The closure failure associated with these displacements is thus

$$\Delta x^k = \Gamma^k_{ml}\, \underset{2}{dx}^m\, \underset{1}{dx}^l + \Gamma^k_{ml}\, \underset{1}{dx}^m\, \underset{2}{dx}^l \tag{5.22a}$$

or more compactly

$$\Delta x^k = 2 S_{lm}^{\cdot\cdot k}\, \underset{1}{dx}^l\, \underset{2}{dx}^m \tag{5.22b}$$

where

$$S_{lm}^{\cdot\cdot k} = \Gamma^k_{[lm]} = \tfrac{1}{2}\left[\Gamma^k_{lm} - \Gamma^k_{ml}\right] \tag{5.23}$$

Thus, we have a somewhat more fundamental definition of the torsion tensor; namely, that it is equal to the asymmetric part of the connection. Furthermore, whereas the connection is not a tensor, the asymmetric part is. It should also be noted that in a Riemannian space (Schouten, 1954)

$$\Gamma^\kappa_{\lambda\mu} = \left\{ {\kappa \atop \lambda\mu} \right\} \tag{5.24}$$

where $\left\{ {\kappa \atop \lambda\mu} \right\}$ are simply Christoffel symbols of the second kind. Since $\Gamma^\kappa_{\lambda\mu}$ in a Riemannian space is symmetric in the indices λ and μ, it follows that $S_{\lambda\mu}^{\cdot\cdot\kappa} = 0$. Thus, all of the elastically distorted spaces considered in section 2 are Riemannian since they do not contain dislocations.

Equation 5.22b could also be written as

$$\Delta x^k = S_{lm}^{\cdot\cdot k}\, dF^{lm} \tag{5.25}$$

or in integrated form as

$$-b^k = \int \Delta x^k = \int_S S_{lm}^{\cdot\cdot k}\, dF^{lm} \tag{5.26}$$

Thus, the integrated value of the covariant differential simply gives the negative of the closure failure associated with a given Burgers circuit. In the case of Figure 5.7*a*

$$S_{12}^{\cdot\cdot 2} = \left\{ -\tfrac{1}{2}(2)\delta(x^1) \right\}_1 + \left\{ \tfrac{1}{2}\delta(x^1) \right\}_2 = \left\{ -\tfrac{1}{2}\delta(x^1) \right\}_1 \tag{5.27}$$

which is the same expression as that given by Eq. 4.78. A similar analysis as that which led to Eq. 5.26 can also be carried out for Figure 5.7*b* in which case we obtain

$$-b^{k^2} = \int \Delta x^{k^2} = \int_s S_{l^2 m^2}^{\cdot\,\cdot\, k^2} dF^{l^2 m^2} \tag{5.28}$$

where

$$S_{12}^{\cdot\,\cdot\,1} = \left\{ -\left(\tfrac{1}{2}\right)\delta(x^1)\tan\theta \right\}_1 + \{0\}_2 \tag{5.29}$$

which is the same as that given by Eq. 4.98.

Now in Figure 5.7, we may write $\underset{1}{dx}^k \equiv dx^1$, $\underset{2}{dx}^k \equiv dx^2$, $\underset{1}{dx}^{k^2} \equiv dx^1$, and $\underset{2}{dx}^{k^2} \equiv dx^2$. Thus, in the case of Figure 5.7*b*, parallel displacement of dx^1 along dx^2 leaves it unchanged so that $\Gamma_{21}^1 = 0$. On the other hand, parallel displacement of dx^2 along dx^1 yields $\Gamma_{12}^1 \neq 0$ or more specifically, in view of Eq. 5.23

$$\Gamma_{12}^1 = 2S_{12}^{\cdot\,\cdot\,1} = \{-\delta(x^1)\tan\theta\}_1 + \{0\}_2 \tag{5.30}$$

In terms of Eq. 5.22a, the above relation yields

$$\Delta x^1 = \Gamma_{12}^1 dx^1 dx^2 = \{-\delta(x^1)\tan\theta\, dx^1 dx^2\}_1 + \{0\}_2 \tag{5.31}$$

which is the dotted closure failure shown in Figure 5.7*b*. Eq. 5.31 in its more general form may be written as (Kröner, 1959)

$$dC^k = -\Gamma_{ml}^k C^m dx^l \tag{5.32}$$

which gives the change in length of a vector C^m as it is displaced in a parallel manner along dx^l.

In the case of Figure 5.7*a*, we have $\Gamma_{12}^2 = 0$ since the vector dx^2 is left unchanged by parallel displacement along dx^1. However, the vector dx^1 is changed by parallel displacement along dx^2. In particular

$$\Gamma_{21}^2 = -2S_{12}^{\cdot\,\cdot\,2} = \{\delta(x^1)\}_1 \tag{5.33}$$

In terms of Eq. 5.22a, the above relation yields

$$\Delta x^2 = \Gamma_{21}^2 dx^2 dx^1 = \{\delta(x^1) dx^2 dx^1\}_1 \tag{5.34}$$

which corresponds to the dotted closure failure shown in Figure 5.7*a*.

Let us now convert the tensor quantities associated with the (k) state into anholonomic coordinates. As noted earlier, such coordinates are designated by lowercase Latin letters that occupy the lower end of the scale, that is, a, b, c, and so forth. The torsion tensor in this system thus becomes (Eringen, 1966)

$$S_{cb}^{\cdot\cdot a} = C_k^a C_c^m C_b^l S_{ml}^{\cdot\cdot k} \tag{5.35}$$

where

$$C_b^l = \delta_b^K \delta_{k'}^l A_K^{k'} \tag{5.36}$$

and where $A_K^{k'}$ is given by Eq. 4.11. Under these conditions, we can write

$$\underset{a}{S_{12}^{\cdot\cdot 2}} = C_2^2 C_1^1 C_2^2 S_{12}^{\cdot\cdot 2} + C_2^2 C_1^2 C_2^1 S_{21}^{\cdot\cdot 2} = \underset{k}{S_{12}^{\cdot\cdot 2}} \tag{5.37}$$

We can also define another quantity $\Omega_{cb}^{\cdot\cdot a}$ termed the object of anholonomity as follows (Schouten, 1954; Zorawski, 1967):

$$\Omega_{cb}^{\cdot\cdot a} \overset{\text{def}}{=} C_c^m C_b^l \partial_{[m} C_{l]}^a \tag{5.38}$$

It is important to note that $\Omega_{cb}^{\cdot\cdot a}$ is not a tensor since a tensor whose components are finite in one system cannot all vanish in another system. In particular, $\Omega_{ml}^{\cdot\cdot k} = 0$. The only nonvanishing component of $\Omega_{cb}^{\cdot\cdot a}$ is

$$\Omega_{12}^{\cdot\cdot 2} = \tfrac{1}{2} \overline{C}_1^1 \overline{C}_2^2 \partial_1 C_2^2 \tag{5.39a}$$

which in view of Eq. 5.36 reduces to

$$\Omega_{12}^{\cdot\cdot 2} = \left\{ -\left(\tfrac{1}{2}\right) 2\delta(x^1) \right\}_1 + \left\{ \left(\tfrac{1}{2}\right) \delta(x^1) \right\}_2 = \left\{ -\tfrac{1}{2} \delta(x^1) \right\}_1 \tag{5.39b}$$

and which in view of Eq. 4.78 means that

$$\Omega_{12}^{\cdot\cdot 2} = S_{12}^{\cdot\cdot 2} \tag{5.40}$$

Now in general, the connection may be written as (Schouten, 1954)

$$\Gamma_{ml}^k = \left\{ {k \atop ml} \right\} + S_{ml}^{\cdot\cdot k} - S_{l\cdot m}^{\cdot k} + S^k_{\cdot ml} - \Omega_{ml}^{\cdot\cdot k} + \Omega_{l\cdot m}^{\cdot k} - \Omega^k_{\cdot ml} \tag{5.41}$$

where quantities such as $S_{l\cdot m}^{\cdot k}$ can be obtained from $S_{ml}^{\cdot\cdot k}$ by employing the metric tensor as follows:

$$S_{l\cdot m}^{\cdot k} = g^{ko} g_{mn} S_{lo}^{\cdot\cdot n} \tag{5.42}$$

Equation 5.41 can aso be expressed in terms of anholonomic coordinates and used to obtain (Schouten, 1954; Eringen, 1966)

$$\Gamma^{a}_{[cb]} = S_{cb}^{\cdot\cdot a} - \Omega_{cb}^{\cdot\cdot a} \tag{5.43}$$

This equation may in turn be used to write the closure failure associated with a given Burgers circuit as

$$b^{a} = -\int_{s} \Gamma^{a}_{[cb]}\, dF^{cb} = -\int_{s} (S_{cb}^{\cdot\cdot a} - \Omega_{cb}^{\cdot\cdot a})\, dF^{cb} \tag{5.44}$$

In holonomic coordinates, the anholonomic object vanishes so that Eq. 5.44 becomes identical in form to that given by Eq. 4.61. In view of Eqs. 5.39 and 5.40, Eq. 5.44 yields

$$b^{a} = b^{a=2} + \underset{s}{b}^{a=2} = (4) + (-4) \tag{5.45}$$

The first term in the sum of the above equation obtains from the $S_{cb}^{\cdot\cdot a}$ term in Eq. 5.44 while the second term in this equation arises from the $\Omega_{cb}^{\cdot\cdot a}$ term in Eq. 5.44. Equation 5.45 is also seen to be equivalent to the sums of Eqs. 5.3a and 5.3b. Thus, we have the interesting result that whereas the torsion tensor is associated with dislocations within a body, the anholonomic object is associated with the free surface or ledge contribution associated with this dislocation. The representation of the (k) state in terms of (a) state coordinates is termed a naturalization process (Kondo, 1955; Kröner, 1958) and is seen to remove all of the elastic strains associated with the distortion. In other words, the naturalization process may be viewed as a process that brings the crystal back into its original Euclidian frame; at least locally.

The (k^2) state can also be represented in terms of anoholonomic coordinates (a^2) in which case we obtain

$$S_{c^2b^2}^{\cdot\cdot a^2} = C_{k^2}^{a^2} C_{c^2}^{m^2} C_{b^2}^{l^2} S_{m^2l^2}^{\cdot\cdot k^2} \tag{5.46}$$

where

$$C_{b^2}^{l^2} = \delta_{b^2}^{K} \delta_{k^{2'}}^{l^2} A_{K}^{k^{2'}} \tag{5.47}$$

and where $A_{K}^{k^{2'}}$ is given by Eq. 4.18. We can thus write

$$\underset{a^2}{S}_{12}^{\cdot\cdot 1} = C_1^1 C_1^1 C_2^2 S_{12}^{\cdot\cdot 1} + C_1^1 C_1^2 C_2^1 S_{21}^{\cdot\cdot 1} = \underset{a^2}{S}_{12}^{\cdot\cdot 1} \tag{5.48}$$

whereas since

$$\Omega_{c^2b^2}^{\cdot\cdot a^2} = C_{c^2}^{m^2} C_{b^2}^{l^2} \partial_{[m^2} C_{l^2]}^{a^2} \tag{5.49}$$

the only nonvanishing component becomes

$$\Omega_{12}^{\cdot\cdot 1} = \tfrac{1}{2} \partial_1 C_2^1 \left[\bar{C}_1^1 \bar{C}_2^2 - \bar{C}_1^2 \bar{C}_2^1 \right] \tag{5.50}$$

In view of Eq. 5.47, the above equation becomes

$$\Omega_{12}^{\cdot\cdot 1} = \left\{ -\left(\tfrac{1}{2}\right) \delta(x^1) \tan\theta \right\}_1 \tag{5.51}$$

Comparing this result with that of Eq. 4.98 we obtain

$$\Omega_{12}^{\cdot\cdot 1} = S_{12}^{\cdot\cdot 1} \tag{5.52}$$

Since

$$b^{a^2} = -\int_S \left(S_{c^2b^2}^{\cdot\cdot a^2} - \Omega_{c^2b^2}^{\cdot\cdot a^2} \right) dF^{c^2b^2} \tag{5.53}$$

Utilizing Eqs. 5.51 and 5.52, Eq. 5.53 yields

$$b^{a^2} = b^{a^2=1} + \underset{S}{b}^{a^2=1} = (4\tan\theta) + (-4\tan\theta) \tag{5.54}$$

which is the same result as Eqs. 5.11a and 5.11b combined. The first term in the above equation arises from the torsion tensor, while the second is due to the anholonomic object. We have thus established a strong physical insight with respect to these two quantities.

The anholonomic states discussed above can also be described in terms of parallel displacements similar to those given for the holonomic states shown in Figure 5.7. In particular, these are shown in Figure 5.8. Similar to Eq. 5.22a, the closure failure associated with the parallel displacement of the vectors $\underset{1}{dx}^a$ and $\underset{2}{dx}^a$ can be written as

$$\Delta x^a = \Gamma_{cb}^a \underset{2}{dx}^c \underset{1}{dx}^b + \Gamma_{cb}^a \underset{1}{dx}^c \underset{2}{dx}^b \tag{5.55a}$$

or equivalently

$$\Delta x^a = 2\Gamma_{[cb]}^a \underset{1}{dx}^c \underset{2}{dx}^b \tag{5.55b}$$

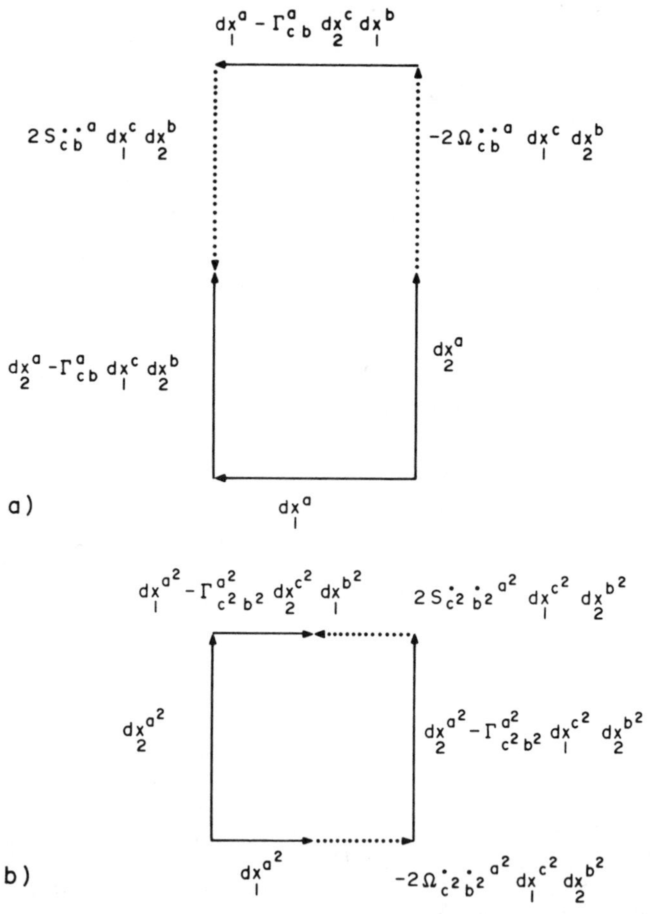

Figure 5.8 Parallel displacement is similar to that shown in Figure 5.7, but expressed in terms of anholonomic coordinates.

where in view of Eq. 5.43

$$\Delta x^a = 2(S_{cb}^{\cdot\cdot a} - \Omega_{cb}^{\cdot\cdot a})\, \underset{1}{dx}^c\, \underset{2}{dx}^b \tag{5.55c}$$

A similar analysis to that above could also be made for the (a^2) state configuration shown in Figure 5.8*b*.

As in the case of Figure 5.7*a*, $\Gamma^2_{12}=0$ for the (a) state of Figure 5.8*a* since the vector dx^2 is unchanged by a parallel displacement along dx^1. On the other hand, dx^1 is changed by parallel displacement along dx^2 so that

$$\Gamma_{21}^{2} = -2(S_{12}^{\cdot\cdot 2} - \Omega_{12}^{\cdot\cdot 2}) = \{\delta(x^1) - \delta(x^1)\}_1 \tag{5.56}$$

or in terms of Eq. 5.55a

$$\Delta x^2 = \Gamma_{21}^{2} dx^2 dx^1 = \{[\delta(x^1) - \delta(x^1)]\, dx^2 dx^1\}_1 \tag{5.57}$$

which corresponds to the pair of dotted closure failures shown in Figure 5.8*a*. A similar analysis can be carried out in connection with Figure 5.8*b*.

We have thus far considered states possessing both a torsion tensor as well as those possessing both torsion as well as an anholonomic object. Next it is of interest to consider states that possess only an anholonomic object. This is best done by referring to the (K) state reference crystal shown in Figure 5.9*a*. This crystal may be pulled apart or torn along the line 3′–4′ or equivalently 2′–1′ to generate the (a^7) state shown in Figure 5.9*b*. The distortion tensor which accomplishes this may be written as $A_K^{a^7}$ and has nonvanishing components

$$A_1^1 = A_3^3 = 1 \tag{5.58a}$$

and

$$A_2^2 = \{H(-x^1)\}_1 + \{H(+x^1)\}_2 \tag{5.58b}$$

For simplicity, the rightmost half of the crystal may be moved to infinity in which case the (a^8) state configuration shown in Figure 5.9*c* obtains. In this case, the distortion tensor $A_K^{a^8}$ has nonvanishing components

$$A_1^1 = A_3^3 = 1 \tag{5.59a}$$

and

$$A_2^2 = \{H(-x^1)\}_1 \tag{5.59b}$$

It follows that the Burgers circuit 4–4′–3′–3–4 in Figure 5.9*b* has a closure failure 3′–4′ associated with it that may be written as

$$b^{a^8} = \oint A_K^{a^8} dx^K \tag{5.60}$$

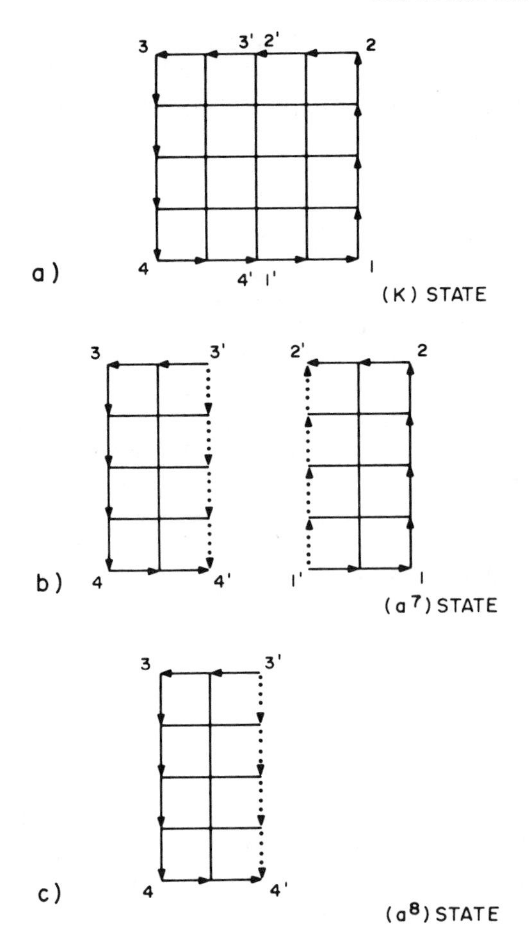

Figure 5.9 *a*) (*K*) state. *b*) and *c*) Torn modifications of the (*K*) state shown in *a*).

or in expanded form as

$$b^{a^8=2}=\int_1^2 A_2^2\,dx^2+\int_2^3 A_1^2\,dx^1+\int_3^4 A_2^2\,dx^2+\int_4^1 A_1^2\,dx^1 \tag{5.61a}$$

which in view of Eq. 5.59 becomes

$$b^{a^8=2}=\int_3^4 A_2^2\,dx^2=\int_3^4 dx^2=-4\equiv \underset{3'-4'}{\Delta x^2} \tag{5.61b}$$

which corresponds to the four dotted arrows along 3′–4′ in Figure 5.9*c*.

Using Stokes' theorem, we can convert the line integral of Eq. 5.60 into a surface integral as follows:

$$b^{a^8}=\oint A_K^{a^8}\,dx^K=\int_s \partial_{[L}A_{K]}^{a^8}\,dF^{LK} \tag{5.62}$$

Integrating with respect to (a^8) state coordinates, the above equation becomes

$$b^{a^8}=\int_s A_{c^8}^{L}A_{b^8}^{K}\partial_{[L}A_{K]}^{a^8}\,dF^{c^8b^8} \tag{5.63}$$

or more compactly as

$$b^{a^8}=\int_s \Omega_{c^8b^8}^{\cdot\;\cdot\;a^8}\,dF^{c^8b^8} \tag{5.64}$$

where

$$\Omega_{c^8b^8}^{\cdot\;\cdot\;a^8}=A_{c^8}^{L}A_{b^8}^{K}\partial_{[L}A_{K]}^{a^8} \tag{5.65}$$

and is just the definition of the anholonomic object. In view of Eq. 5.59, it is easy to show that the only nonvanishing component of $\Omega_{c^8b^8}^{\cdot\;\cdot\;a^8}$ is

$$\Omega_{12}^{\cdot\;\cdot\;2}=\tfrac{1}{2}\bar{A}_1^1\bar{A}_2^2\partial_1 A_2^2 \tag{5.66}$$

or

$$\Omega_{12}^{\cdot\;\cdot\;2}=-\tfrac{1}{2}\delta(x^1) \tag{5.67}$$

Substituting this result into Eq. 5.64 gives

$$b^{a^8=2}=\int_s \Omega_{12}^{\cdot\;\cdot\;2}\,dF^{12}=-4 \tag{5.68}$$

which is the same result as that given by the line integral procedure that led to Eq. 5.61b. It is important to note that there are no dislocations associated with the (a^8) state since there is no torsion tensor associated with this state. This is easy to show since similar to Eq. 5.46

$$S_{c^8b^8}^{\cdot\;\cdot\;a^8}=C_K^{a^8}C_{c^8}^{M}C_{b^8}^{L}S_{ML}^{\cdot\;\cdot\;K} \tag{5.69}$$

where

$$C_K^{a^8} \equiv A_K^{a^8} \tag{5.70}$$

where $A_K^{a^8}$ is given by Eq. 5.59. Since $S_{ML}^{\cdot\cdot K} = 0$, it follows that $S_{c^8 b^8}^{\cdot\cdot a^8} = 0$.
On the other hand,

$$\Omega_{c^8 b^8}^{\cdot\cdot a^8} = C_{c^8}^M C_{b^8}^L \partial_{[M} C_{L]}^{a^8} \tag{5.71}$$

which is the same as that given by Eq. 5.65. We thus conclude that the (a^8) state possesses only an anholonomic object and that this anholonomic object is not a tensor since it transforms according to (Gołąb, 1974)

$$\Omega_{c'b'}^{\cdot\cdot a'} = C_{c'}^c C_{b'}^b C_a^{a'} \Omega_{cb}^{\cdot\cdot a} - C_{[c'}^a \partial_{b']} C_a^{a'} \tag{5.72}$$

Similar to Eq. 5.23, it is possible to write the anholonomic object for the (a^8) state as follows:

$$-\Omega_{c^8 b^8}^{\cdot\cdot a^8} = \Gamma_{[c^8 b^8]}^{a^8} = \tfrac{1}{2}\left[\Gamma_{c^8 b^8}^{a^8} - \Gamma_{b^8 c^8}^{a^8}\right] \tag{5.73}$$

Referring now to Figure 5.10, which is the analogue of Figure 5.9*c*, described in terms of parallel transport, it is apparent that since the vector dx^2 is left unchanged by parallel displacent along dx^2, it follows that $\Gamma_{12}^2 = 0$. On the other hand, the vector dx^2 is reduced to zero upon parallel transport along dx^1. Eq. 5.73 thus gives

$$\Gamma_{21}^2 = 2\Omega_{12}^{\cdot\cdot 2} = -\delta(x^1) \tag{5.74}$$

The change in length of the vector dx^2 upon parallel transport is thus given

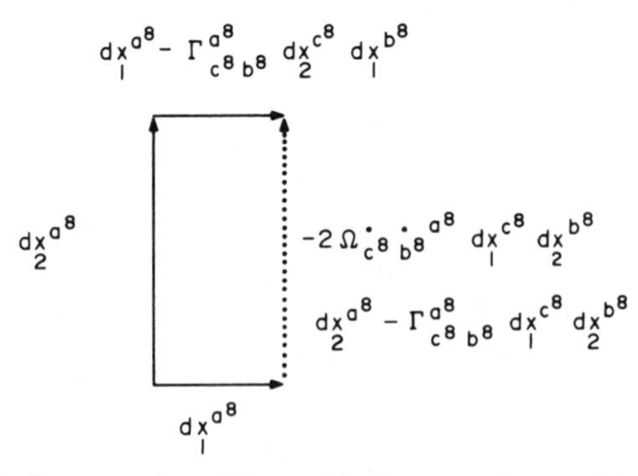

Figure 5.10 Interpretation of Figure 5.9*a* in terms of a parallel displacement.

by

$$\Delta x^2 = \Gamma^2_{21}\, dx^2 dx^1 = \delta(x^1)\, dx^2 dx^1 \tag{5.75}$$

that coincides with the dotted closure failure shown in Figure 5.10.

As in the case of Eq. 4.115, Eq. 5.64 can be written in differential form to yield

$$db^{a^8} = \Omega_{c^8 b^8}^{\cdot\cdot\; a^8}\, dF^{c^8 b^8} = \Omega_{c^8 b^8}^{\cdot\cdot\; a^8}\, \varepsilon^{d^8 c^8 b^8}\, dF_{d^8} = \beta^{d^8 a^8}\, dF_{d^8} \tag{5.76}$$

The quantity $\beta^{d^8 a^8}$ may be viewed as the density of newly created free surface and is given by

$$\beta^{d^8 a^8} = \varepsilon^{d^8 c^8 b^8}\, \Omega_{c^8 b^8}^{\cdot\cdot\; a^8} \tag{5.77}$$

For the (a^8) state of Figure 5.9*c*, it is obviously

$$\beta^{32} = 2\Omega_{12}^{\cdot\cdot\;2} = -\delta(x^1) \tag{5.78}$$

where $\Omega_{12}^{\cdot\cdot\;2}$ in the above expression is given by Eq. 5.67. Upon integration, Eq. 5.78 yields the value of -1, and in terms of Figure 5.9*c*, corresponds to the length 3′–4′ divided by 4–3, that is, the length of newly created free surface divided by the original length of perfect crystal. In the case of the (a) and (a^2) states of Figure 5.8, we obtain

$$\underset{a}{\beta^{32}} = 2\Omega_{12}^{\cdot\cdot\;2} = \{-\delta(x^1)\}_1 \tag{5.79a}$$

and

$$\underset{a}{\beta^{31}} = 2\Omega_{12}^{\cdot\cdot\;1} = \{-\delta(x^1)\tan\theta\}_1 \tag{5.79b}$$

In terms of Figure 5.1*a*, Eq. 5.79a corresponds to the four downward pointing arrows along 2–1 divided by the original length of this line in the reference (K) state crystal. On the other hand, Eq. 5.79b corresponds to the dotted line 1′–1 in Figure 5.2*b* divided by the length 1′–2.

5.2 SURFACE DISLOCATIONS

Figure 5.11*a* shows a crystal containing a single edge dislocation that, except for the absence of the ledge on the leftmost face of the crystal, is similar to that given in Figure 4.4*b*. It is clear that between the vertical segments 1′–2″ and 3″–4′, the surface of the body is not stress free. As is shown in a later section, such a configuration is possible only when the surface tension can support such stresses. For simplicity, however, in this section we consider only those cases of stress free surfaces. These are in fact the boundary conditions used by nearly everyone. The surfaces in Figure 5.11*a* can be made stress free by the introduction onto the surface of tiny dipoles such as shown in Figure 5.11*b*. One such dipole is shown in enlarged form towards the right of this figure. The purpose of the dipole is to act as a link or transition zone between the stressed interior and the stress-free surface. It is also noted that the dipole also provides a segment of newly created free surface that is shown by a dotted arrow. In order to emphasize the uniqueness of these surface dislocation dipoles, one half of the dislocation pair comprising the dipole is drawn solid, while the other is drawn dotted. The dislocation drawn solid provides the stress field, but no free surface, while the dislocation shown dotted possesses no stress field, but has associated with it a free surface. The surface dislocation dipoles may be considered as infinitesimal in strength and uniformly distributed over the surface of the dislocated crystal.

An alternate way of viewing the configuration shown in Figure 5.11*b* is illustrated in Figure 5.11*c*. In particular, the latter configuration may be viewed as an infinite crystal containing a dislocation. Additional dislocations of smaller Burgers vector are then arranged such that the stress vanishes on a given set of planes, namely the planes 1–1′–2″–2–3–3″–4′–4–1. It then becomes possible to cut the body along these planes and remove it from the infinite body so as to obtain the configuration shown in Figure 5.11*b*. Strictly speaking, however, the cutting just described creates all new free surfaces, so that the final configuration is not exactly the same as that shown in Figure 5.11*b*. Finally, when the dislocation in Figure 5.11 moves to the rightmost face of the crystal, the configuration shown in Figure 5.11*d* obtains. In this case, the topmost dislocations associated with all of the surface dipoles annihilate with the internal dislocation and leaves a single dislocation along with its accompanying free surface, both of whose strengths are equal to that of the original internal dislocation. Another important finding is that the sum of the lengths of all the extra free surfaces associated with the ledges in Figure 5.11*b* are equal to the negative of the Burgers vector associated with the internal dislocation. More specifically

$$b+\sum_{s} b=0 \tag{5.80a}$$

This extremely important result was first formulated in connection with surface cracks (Marcinkowski and Das, 1974). Note that when formulated

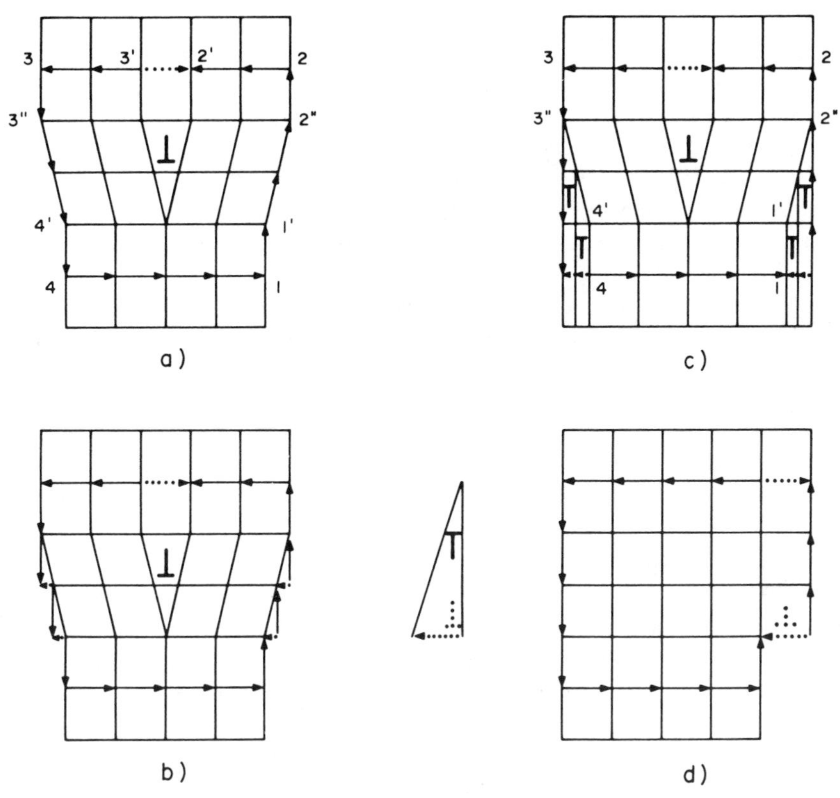

Figure 5.11 Dislocated crystal. *a*) Without surface dislocations. *b*) With surface dislocations. *c*) Alternative representation of Figure 5.11*b* in terms of an infinite crystal. *d*) Configuration attained after the dislocation in Figure 5.11*b* has moved to the rightmost face of the crystal.

in terms of Figure 5.11*c*, Eq. 5.80a becomes

$$\sum b = 0 \tag{5.80b}$$

which simply states that the sum of the Burgers vectors of all the dislocations within the infinite body is zero. This condition is also equivalent to the result that no long-range stresses exist within the infinite body. It also follows from the analysis of the previous section that the ledges in Figure 5.11*b* and their associated dislocations, shown dotted, possess both a torsion tensor and anholomonic object that satisfy a relation of the type given by Eq. 5.52, that is,

$$S_{cb}^{\cdot\cdot a} = \Omega_{cb}^{\cdot\cdot a} \tag{5.81}$$

On the other hand, for the nonledge associated dislocation of the dipole, that is, the topmost dislocation that is drawn solid, as well as the internal dislocation in Figure 5.11*b*, $S_{cb}^{\cdot\cdot a} \neq 0$, where as $\Omega_{cb}^{\cdot\cdot a} = 0$.

It follows that the (k') state crystal shown in Figure 4.2 also has nonvanishing surface stresses. These can be removed by the introduction of surface dislocations in the manner shown in Figure 5.12. In a similar manner, suitable surface dislocation arrays can be introduced so as to eliminate the surface stresses associated with the dislocated configurations illustrated in Figures 4.1, 4.3–4.6.

As the internal dislocation in Figure 5.11*b* moves to the bottom of the crystal, the surface dislocations begin to accumulate in the bottommost surface of the crystal as shown in Figure 5.13*a*. These surface dislocations are best illustrated with respect to Figure 5.13*b*. In particular, the leftmost drawing shows the formation of a *V*-notched surface crack. It is apparent that the formation of such a crack can eliminate the stresses on the bottommost face of the crystal. However, the crack faces are not stress free in the same manner as the faces 1′–2″ and 4′–3″ in Figure 5.11*a* are not stress free. In order to reduce this stress, a set of steps is nucleated on each of these crack surfaces in much the same way as the steps were nucleated in Figure 5.11*b*. This is shown in the center drawing of Figure 5.13*b*. Finally, upon coalescence of the two steps the surface dislocation dipole shown to the right of Figure 5.13*b* obtains. This is the actual nature of the surface dislocations that are shown in Figure 5.13*a*. It is thus seen that in principle these surface dislocations are identical to those shown in Figure 5.11*b*, as is to be expected.

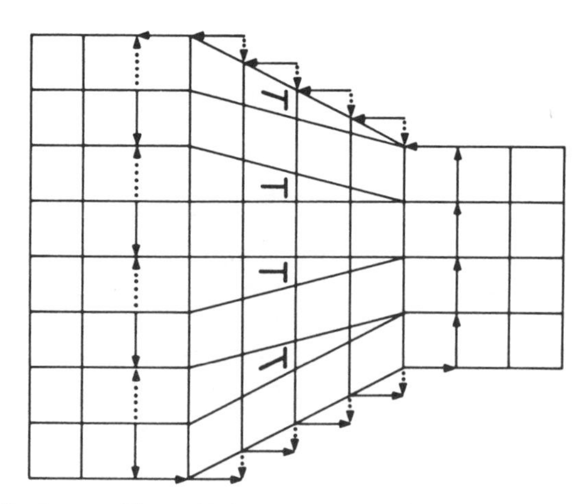

Figure 5.12 Same as Figure 4.2, but with the inclusion of surface dislocations.

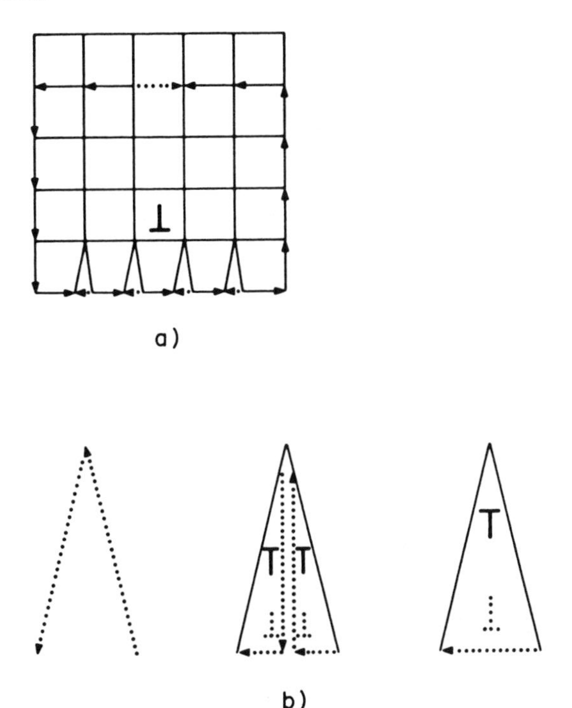

Figure 5.13 *a*) Movement of surface dislocations to bottommost surface of crystal. *b*) Detailed analysis of surface dislocations shown in *a*).

Although the surface dislocations considered thus far are actually dipoles, they have for convenience been referred to simply as surface dislocations. Along these same lines, a somewhat more schematic illustration of Figure 5.11*b* is shown in Figure 5.14*a*. The surface dislocations shown therein simply represent the topmost dislocation of the dipole, and is in fact the one that gives rise to the stress field, since the bottommost dislocation is associated with the stress-free ledge, as discussed earlier.

Strictly speaking, the surface dislocation array shown in Figure 5.14*a* is not complete. This can be seen by referring to Figure 5.14*b* which is a somewhat more detailed drawing of the dislocation configurations shown in Figure 5.11. In particular, the dislocation is also seen to give rise to a bending or warping of the horizontal planes (Jagannadham and Marcinkowski, 1978b; Marcinkowski 1978e). In order to satisfy the stress-free surface boundary conditions, a secondary set of surface dislocations must be added to the primary array whose Burgers vectors are at right angles to those in the former array (Jagannadham and Marcinkowski, 1978b; Marcinkowski 1978e). The topmost dislocations of the secondary

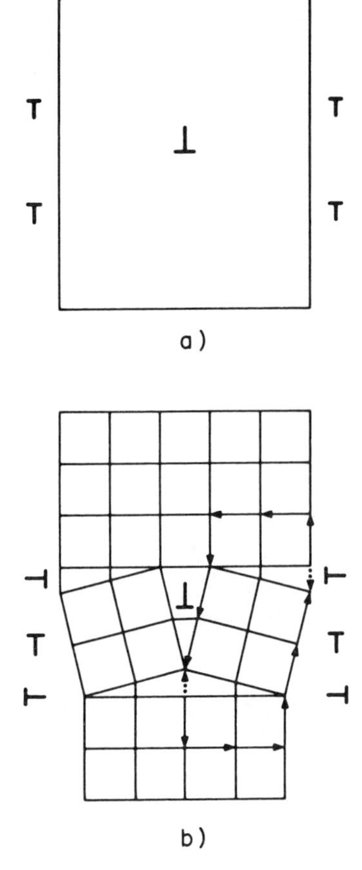

Figure 5.14 *a*) Simplified representation of surface dislocations in Figure 5.11. *b*) Incorporation of secondary surface dislocation array.

array in Figure 5.14*b* are seen to be identical to those associated with the primary array in Figure 5.13. For completeness, a Burgers circuit is taken about the rightmost half of the body in Figure 5.14*b*. The closure failures associated with the secondary array of surface dislocations on the rightmost face of the crystal are shown by dotted arrows. Finally, also for completeness, a secondary array of surface dislocations would have to be added to the bottommost surface in Figure 5.13*a*.

Surface dislocations form not only to satisfy the stress-free boundary conditions arising from internal stresses due to dislocations, but may also be associated with external stresses. As an example, consider the body shown in Figure 5.15*a* which contains a rectangular hole at its center. Upon application of a uniform vertical tensile stress, the hole deforms to the configuration shown in Figure 5.15*b*. In order to satisfy the stress-free boundary conditions on the horizontal and vertical surfaces of the hole, a

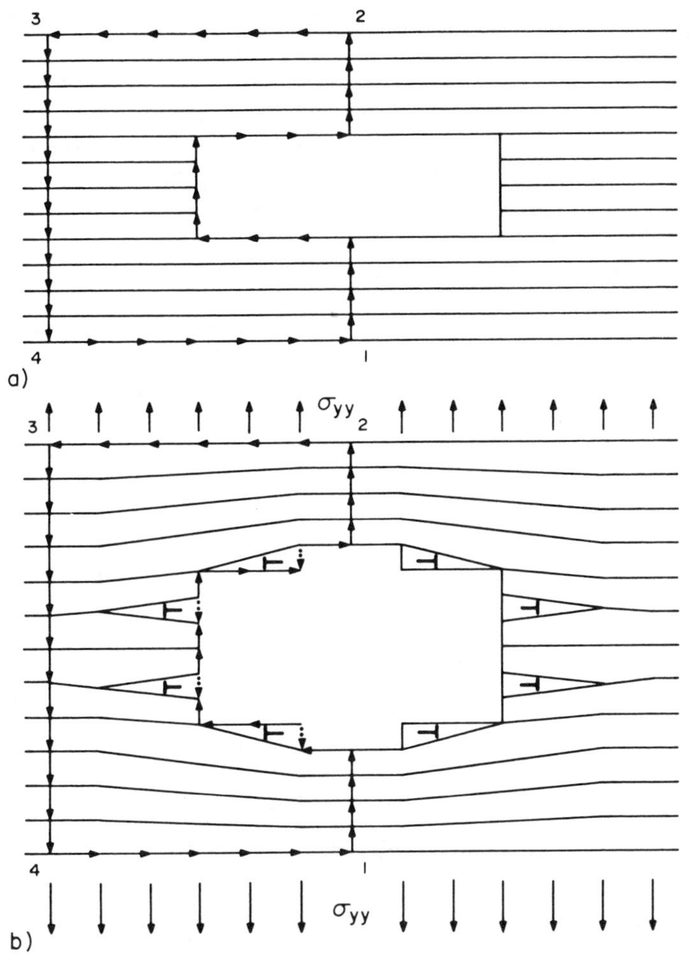

Figure 5.15 *a*) Unstressed body containing a hole. *b*) Dislocation model of this hole after stressing. Reprinted by permission of Springer-Verlag from M. J. Marcinkowski, (1978e), *Acta Mechanica*, in publication, Figure 6a and b.

set of surface dislocations similar to those shown in Figures 5.11*b* and 5.13*a*, respectively, form on these surfaces (Marcinkowski and Das, 1974; Jagannadham and Marcinkowski, 1978a; Marcinkowski, 1972). Burgers circuits constructed with respect to the left half of the hole clearly shown the closure failures associated with these surface dislocations. In the limiting case where the top and bottom faces of the hole approach one another, the hole degenerates into a crack. More is said with respect to this problem later. In the other limiting case where the top and bottom faces of the hole approach infinity two stressed half spaces are generated.

Next let us consider the reference state crystal shown in Figure 5.16*a*. If the region towards the left of the dotted line is subjected to a pure elastic uniaxial tension such as shown in Figure 2.1*b*, while the region towards the right of this dotted line is subject to a pure plastic uniaxial tension such as shown in Figure 3.1*a*, the configuration shown in Figure 5.16*b* obtains. It is apparent that the interface between elastically and plastically strained crystal may be regarded as a vertical array of surface dislocations. It is apparent that the material towards the right of the dotted line is unstressed so that a cut could be made along this dotted line and the unstressed

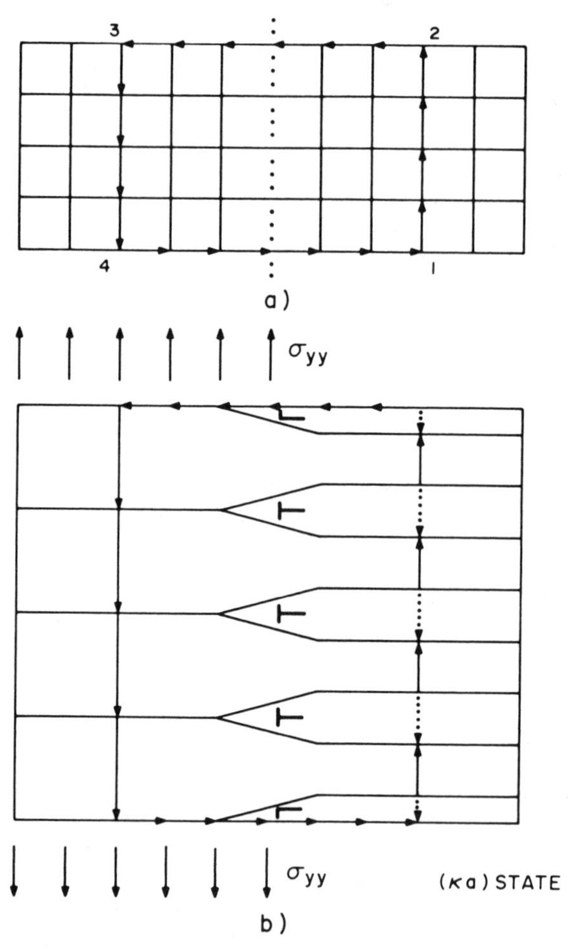

Figure 5.16 *a*) Original reference crystal from which *b*) is formed. *b*) Elastically and plastically strained regions separated by an interface.

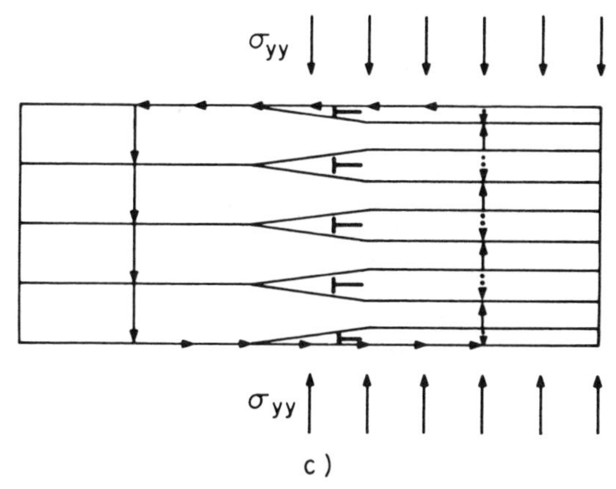

Figure 5.16 (Continued) *c*) Same as Figure 5.16*b* but after application of a reverse external stress.

material removed, so that we obtain the exact configuration shown by the leftmost vertical hole surface in Figure 5.15*b*. Finally, after a stress of equal magnitude, but opposite sign to that shown in Figure 5.16*b* is applied to the horizontal faces of this Figure, the configuration shown in Figure 5.16*c* obtains. In particular, the left half of the body is now stress free, whereas the rightmost half of the body is in a state of simple uniaxial elastic compression similar to that shown in Figure 2.1*c*. It is important to note here that the vertical arrows in Figure 5.16*c* represent the compressive stress generated by the dislocation array shown therein or vice versa. In particular, they are one and the same stress and do not correspond to separate contributions. This becomes more clear shortly. The counterpart of Figure 5.16*c* for a finite body is shown in Figure 5.17. In particular, if the hole in Figure 5.15*b* is filled with stress free matter followed by removal of the applied stress σ_{yy}, then the top and bottom faces of the material within the hole must be subjected to a uniform compressive stress $-\sigma_{yy}$ since in the presence of the applied stress, the stress on these surfaces was zero everywhere. The resulting configuration is shown in Figure 5.17. In other words, the dislocation configuration shown therein describes the uniform compressive stresses on both faces of the finite body. Thus we have the important result that not only can a suitable dislocation array make the stresses vanish on a given surface, but such arrays can be used to describe any given state of externally applied stress. The configuration shown in Figure 5.17 obviously corresponds to a state of simple compression. Strictly speaking, in order to make the vertical surfaces in this figure stress free, a secondary array of dislocations must be added; as shown schematically in Figure 5.18. It is the secondary array that accounts for the

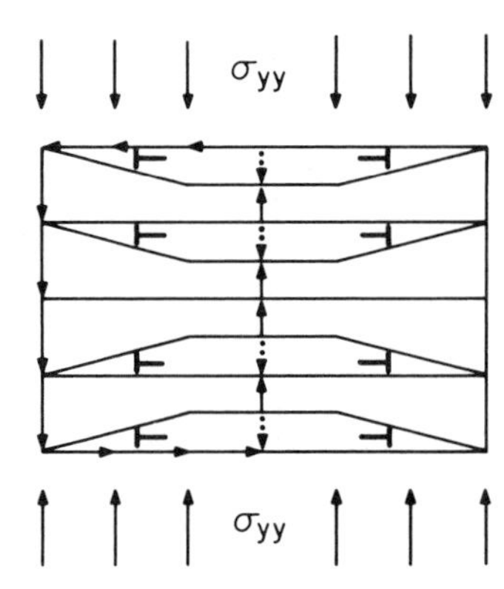

Figure 5.17 Configuration obtained after hole in Figure 5.15*b* is filled with stress-free matter followed by addition of a reverse external stress.

bulging or barreling of the vertical faces of the body. On the other hand, the primary surface dislocation array accounts for the overall contraction of the body upon compression.

The operations described in the previous paragraph can be presented in a somewhat more general form by considering the infinite body subjected to a uniform tensile stress such as shown in Figure 5.19*a*. A rectangular hole may then be cut from the infinite body as shown in Figure 5.19*b*. In order to allow the surface tractions on the faces of the hole to vanish, a set of surface dislocation dipoles is distributed over these faces. For simplicity, only one such dipole is shown in Figure 5.19*b*. A set of dipoles of opposite sense to that associated with the hole may next be distributed over the faces of the finite rectangular piece with the same distribution as that for

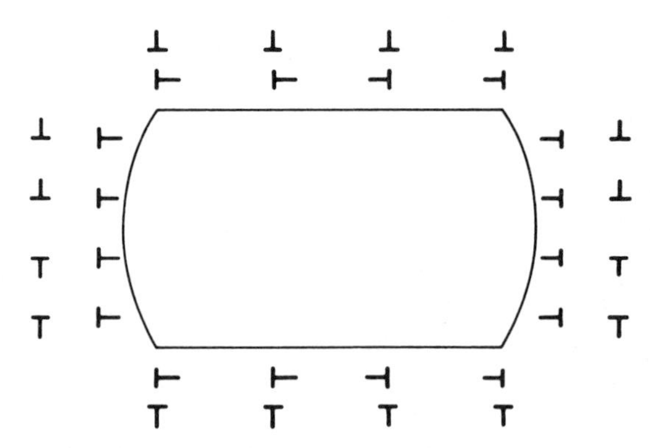

Figure 5.18 More complete description of Figure 5.17 in terms of both primary and secondary surface dislocation arrays. Reprinted by permission of Elsevier Sequoia from K. Jagannadham and M. J. Marcinkowski, (1978c), *Materials Science and Engineering*, in publication, Figure 1hII.

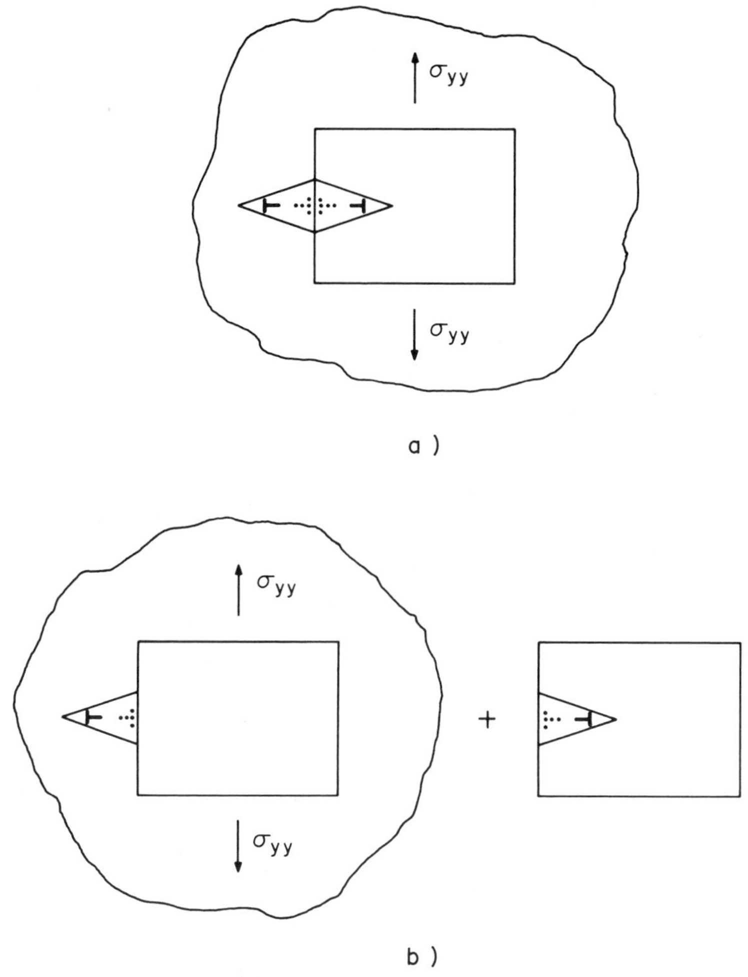

Figure 5.19 *a*) Infinite body subjected to a uniform tensile stress σ_{yy}. *b*) Same body as in *a*) after creation of a hole by removal of matter. Reprinted by permission of Elsevier Sequoia from K. Jagannadham and M. J. Marcinkowski, (1978c), *Materials Science and Engineering*, in publication, Figure 1a and b.

the hole as shown in Figure 5.19*b*. When the two pieces shown in Figure 5.19*b* are reassembled, the two sets of surface dislocation arrays vanish and we regain the uniformly stressed body shown in Figure 5.19*a*.

A set of prismatic dislocations can be introduced next into the rectangular body of Figure 5.19*b* such as shown at the upper right corner of Figure 5.19*c*. The stress field of these dipoles is seen to annihilate with the stress

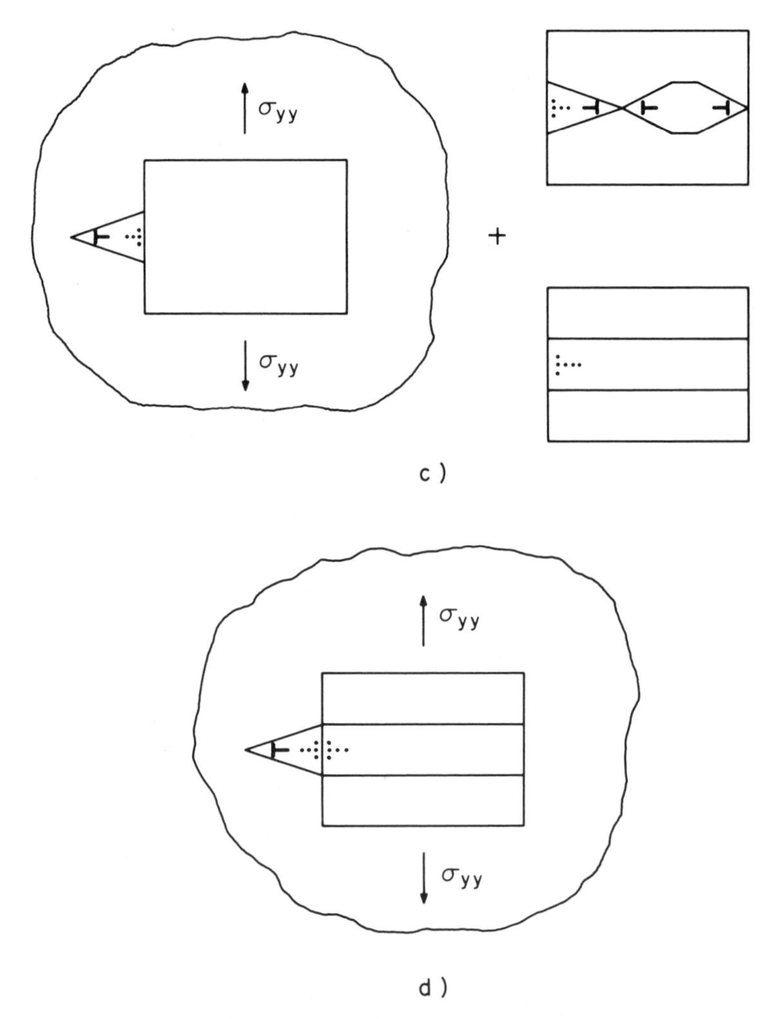

Figure 5.19 (Continued) *c*) Same as in *b*) after addition of prismatic dislocation in removed matter. *d*) Same as in *c*) after insertion of removed matter back into hole. Reprinted by permission of Elsevier Sequoia from K. Jagannadham and M. J. Marcinkowski, (1978c), *Materials Science and Engineering*, in publication, Figure 1c and d.

fields of the surface dislocations to yield the stress-free body shown in the lower right corner of Figure 5.19*c*. If this stress-free body is reinserted into the hole shown in Figure 5.19*c*, the configuration shown in Figure 5.19*d* obtains. A uniform reverse, i.e., compressive stress of magnitude equal to that of the original tensile stress may now be applied to the body in Figure 5.19*d* to obtain the configuration shown in Figure 5.19*e*. The rectangular

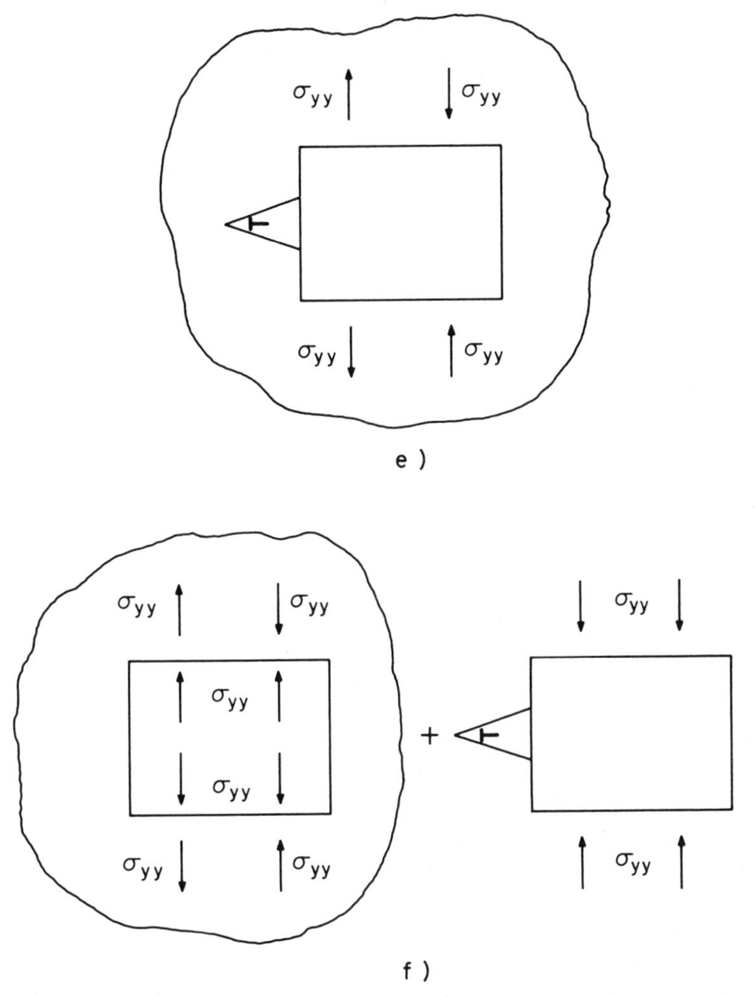

Figure 5.19 (Continued) *e*) Same as in *d*) after application of a reverse uniform tensile stress σ_{yy} of same magnitude as that employed in *a*). *f*) Same as in *e*) after removal of matter to create hole. Reprinted by permission of Elsevier Sequoia from K. Jagannadham and M. J. Marcinkowski, (1978c), *Materials Science and Engineering*, in publication, Figure 1e and f.

body in Figure 5.19*e* can again be cut from the infinite body such as shown in Figure 5.19*f*. However, in order to maintain equilibrium, a set of external stresses equal in magnitude to σ_{yy} must be placed on both horizontal faces of the hole, as shown in Figure 5.19*f*. It is clear that the rectangular body in Figure 5.19*f* corresponds to that shown in Figure 5.17.

When the hole shown in Figure 5.15*a* is subjected to a uniform compressive stress, the configuration shown in Figure 5.20 obtains. Again, as in

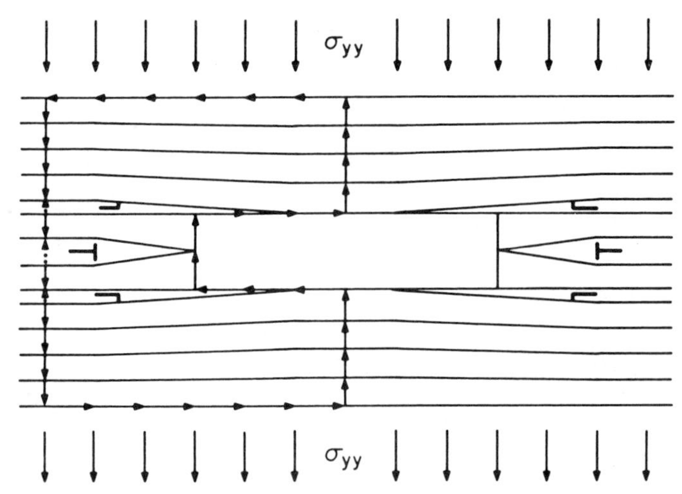

Figure 5.20 Application of a uniform compressive stress to the hole shown in Figure 5.15*a*.

Figure 5.15*b*, the hole must be covered with surface dislocations in order to satisfy the stress-free boundary conditions; however, in comparison to the case in Figure 5.15*b*, the surface dislocations in Figure 5.20 are of opposite sign.

If the reference body shown in Figure 5.16*a* is subjected to an elastic compressive strain towards the left of the dotted line similar to that given in Figure 2.1*c*, while the material to the right of this line is strained plastically by the same amount similar to that shown in Figure 3.1*b*, the configuration shown in Figure 5.21*a* obtains. The dislocations that separate the elastic and plastic regions are seen to be formally equivalent to the surface dislocations on the vertical faces of the hole in Figure 5.20. If a reverse stress of equal magnitude is applied to the body in Figure 5.21*a*, the configuration shown in Figure 5.21*b* obtains. The leftmost side of the body is now stress free, whereas the rightmost half of the body is under a state of uniform tension.

The arguments considered in the previous paragraph can be applied to a finite body by first considering Figure 5.20. If the material within the hole is filled with stress-free material, and the external compressive stress removed, then the configuration in Figure 5.22*a* obtains. In particular, the top and bottom faces of the hole become subject to a uniform tensile stress of magnitude equal to that of the original compressive stress. A more exact description of Figure 5.22*a* is shown in Figure 5.22*b* where the secondary dislocation array has also been included. It is noted that the secondary array of dislocations on the vertical faces of the body gives rise to a concave bending of these surfaces.

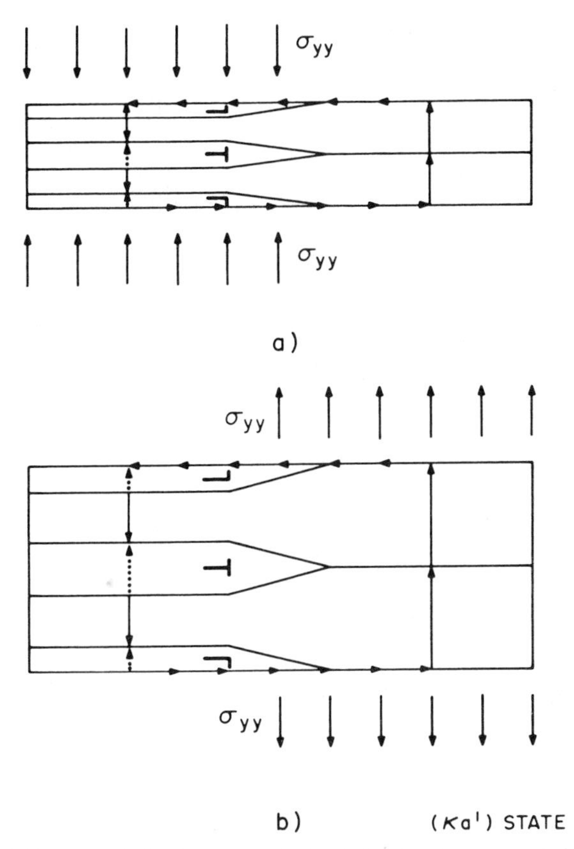

Figure 5.21 *a*) Elastic and plastic compressively strained regions separated by an interface. *b*) Same as *a*), but after application of a reverse external stress.

The body containing the hole in Figure 5.15*a* can also be subjected to a simple shear stress that leads to the configuration shown in Figure 5.23. In order to make the vertical walls of the hole stress free, surface dislocations form on these surfaces as shown. These surface dislocations are also similar to the interface dislocations shown in Figure 5.24*a* that form when the leftmost portion of the reference body shown in Figure 5.16*a* is subjected to a simple elastic shear such as shown in Figure 2.2*a*, while the rightmost half of the body is subjected to a simple plastic shear such as shown in Figure 3.2*a*. If a reverse shear stress of equal magnitude is applied to the body in Figure 5.24*a*, the configuration shown in Figure 5.24*b* obtains. The rightmost half of Figure 5.24*b* is similar to what obtains when the hole in Figure 5.23 is filled with stress-free matter followed by the application of reverse shear stress of equal magnitude. The resultant

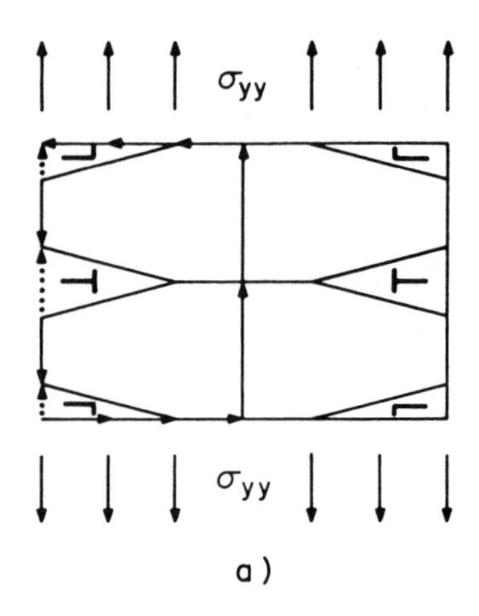

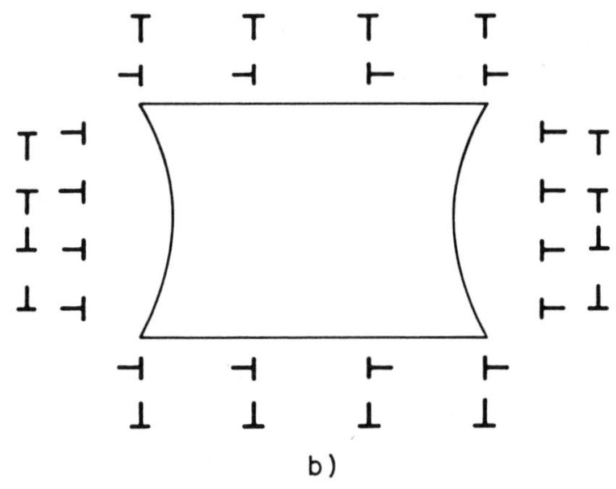

Figure 5.22 *a*) Configuration obtained after hole in Figure 5.20 is filled with stress-free matter followed by addition of a reverse external stress. Reprinted by permission of Elsevier Sequoia from K. Jagannadham and M. J. Marcinkowski, (1978c), *Materials Science and Engineering*, in publication, Figure 1iII.

configuration is shown in Figure 5.25*a* and corresponds to a finite stressed body subject to a uniform shear stress on both horizontal faces. A more complete description of Figure 5.25*a* is given in Figure 5.25*b* where both primary and secondary dislocation arrays are included. Note that the secondary dislocation arrays on the vertical faces give rise to the *S* shape appearance of these surfaces, whereas the primary array accounts for the overall rotation of the vertical planes within the body. For convenience,

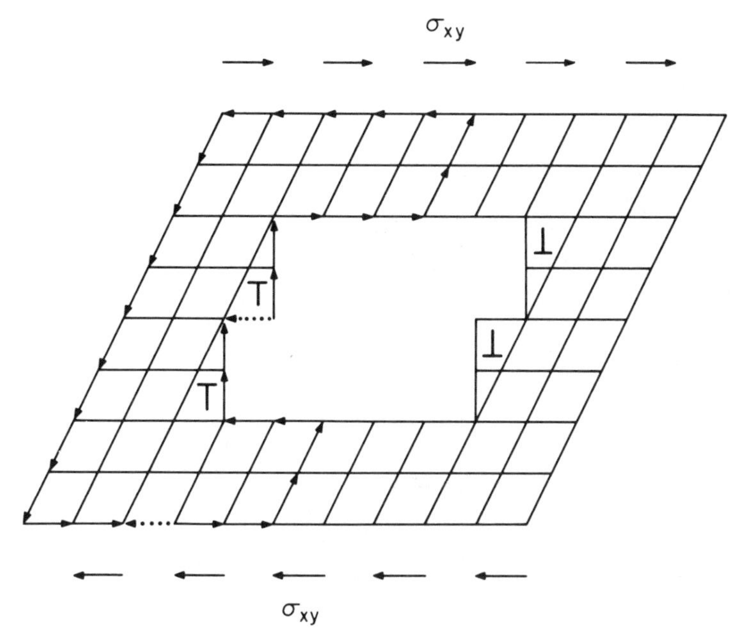

Figure 5.23 Application of a uniform simple shear stress to the hole shown in Figure 5.15*a*.

the dislocations shown in Figure 5.25*b* are drawn with orthogonal Burgers vectors. It is shown later that this would in fact be the correct representation in the limit of small distortions, that is the linear approximation.

If the body containing the hole in Figure 5.15*a* is subjected to a double shear, the configuration shown in Figure 5.26 obtains. All four surfaces of the hole can be made stress free by the addition of surface dislocations as shown in the figure. The set of surface dislocations on the leftmost face of the hole is similar to the interface dislocations shown in Figure 5.27*a* which are generated when the leftmost portion of the reference body shown in Figure 5.16*a* is subject to a double elastic shear such as shown in Figure 2.3*a*, while the rightmost portion of the body is subject to a double plastic shear such as shown in Figure 3.3*a*. Upon application of a reverse double shear stress to the body shown in Figure 5.27*a*, the configuration shown in Figure 5.27*b* obtains. The rightmost configuration in Figure 5.27 is similar to that which obtains when the hole in Figure 5.26 is filled with stress-free material followed by application of a reverse double shear stress of equal magnitude. The resultant configuration is shown in Figure 5.28*a* and corresponds to a finite stressed body subject to a set of shear stresses on all four faces of the body. A more complete description of Figure 5.28*a*

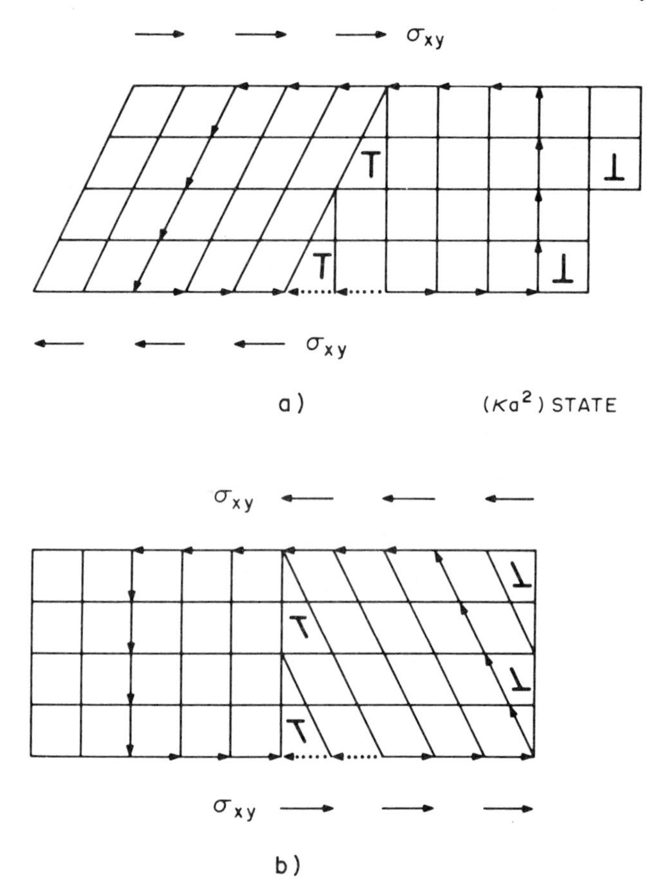

Figure 5.24 *a*) Elastically and plastically sheared regions separated by an interface. *b*) Same as *a*), but after application of a reverse external stress.

is shown in Figure 5.28*b* where both primary and secondary arrays are included. The primary arrays are the inner ones.

The body containing the hole in Figure 5.15*a* could also be rigidly rotated in which case a configuration similar to that shown in Figure 5.26 also obtains. The exact nature of these surface dislocations can be better understood by reference to Figure 5.29, where the leftmost portion of the body is subject to a rigid elastic rotation, similar to that shown in Figure 2.2*b*, while the rightmost half of the body is subject to a plastic rotation such as shown in Figure 3.2*c*. It is noted that in contrast to Figure 5.27*a*, two sets of primary dislocations are required at the interface. Upon application of a reverse rotation to the body illustrated in Figure 5.29, the

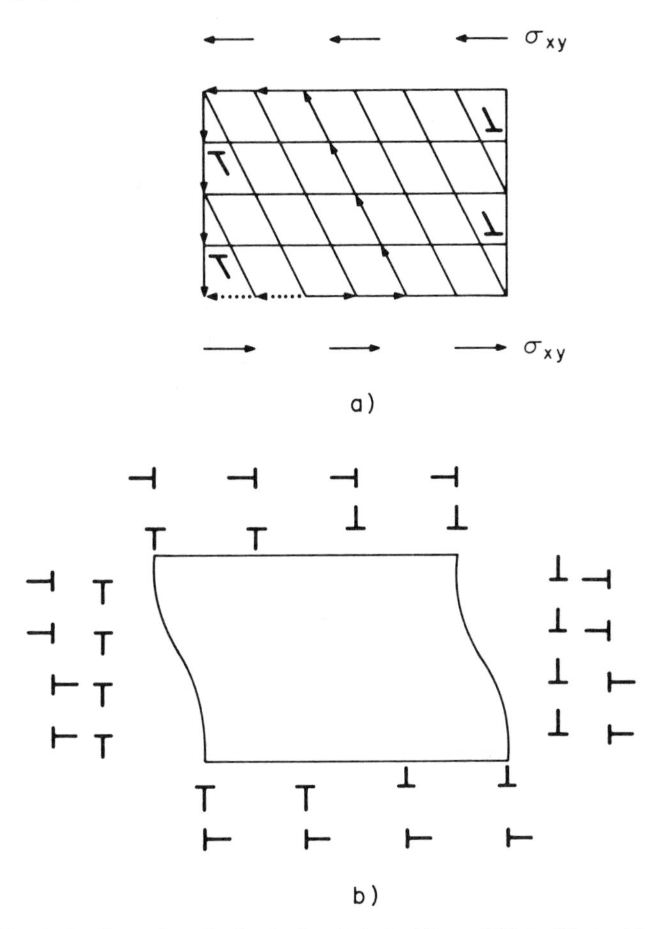

Figure 5.25 *a*) Configuration obtained after hole in Figure 5.23 is filled with stress-free matter followed by addition of a reverse external stress. Reprinted by permission of Elsevier Sequoia from K. Jagannadham and M. J. Marcinkowski, (1978c), *Materials Science and Engineering*, in publication, Figure 4b.

configuration shown in Figure 5.29*b* obtains. The rightmost portion of Figure 5.29*b* is similar to what one obtains when the hole in Figure 5.26 is filled with stress-free material followed by a reverse rotation of equal magnitude. The resultant configuration is shown in Figure 5.30. Note that there are now no external stresses on the body, that is to be expected for rigid rotations. Note also that unlike Figure 5.28*a*, Figure 5.30 contains two sets of dislocations. The ones shown outside the body compensate for the volume increase associated with the dislocations inside the body. Also

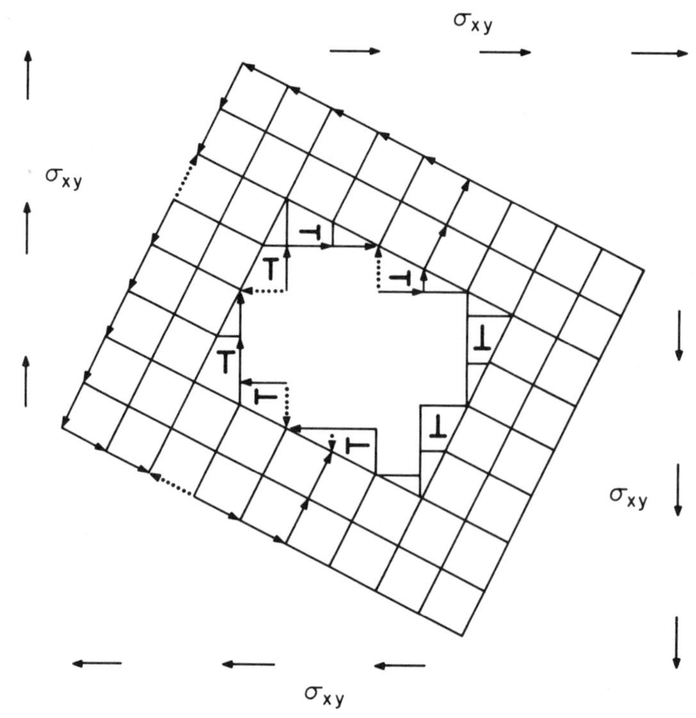

Figure 5.26 Application of a double shear stress to the hole shown in Figure 5.15*a*.

it is important to note that both sets of dislocations shown in Figure 5.30 are of primary type. Since there is no stress associated with the finite body of Figure 5.30, unlike the case shown in Figure 5.28*b*, there is no change in shape of the outer surfaces, so that no secondary dislocations, that account for this shape change, are required.

Finally, the plastically and elastically strained states shown in Figures 3.4 and 2.4, respectively, that consist of pure shears, can be used to obtain the surface dislocation array shown in Figure 5.31*a*. In particular, it denotes the surface dislocation analogue of a finite body subjected to a pure shear. The more complete description which includes the incorporation of the secondary surface dislocation array is shown in Figure 5.31*b* where once again, the secondary array accounts for the change in shape of the external surfaces.

Thus it follows from the discussions in the preceding paragraphs that any set of boundary conditions, either stressed or stress free, can be represented in terms of some suitable arrangement of dislocations on these surfaces (Jagannadham and Marcinkowski, 1978c). This is an extremely

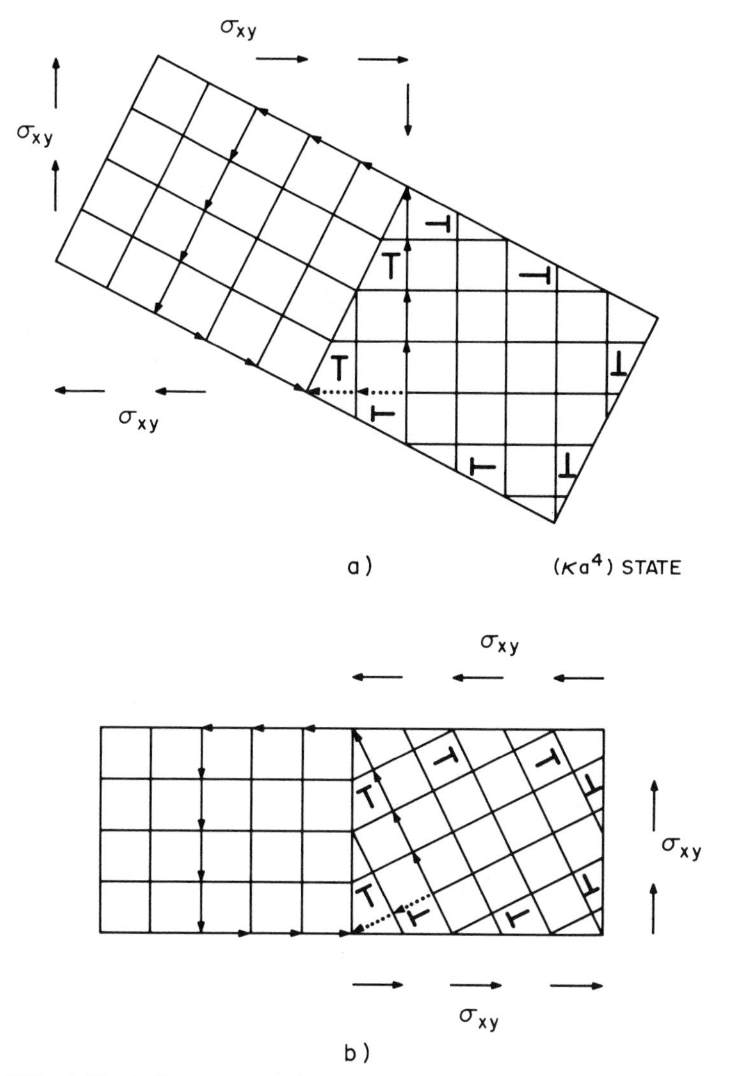

Figure 5.27 *a*) Elastically and plastically double sheared regions separated by an interface. *b*) Same as *a*), but after application of a reverse external stress.

important result and allows any problem with the realm of elasticity theory to be formulated in terms of surface dislocations. More is said with respect to this important point in a later section.

Let us now return to Figure 5.16*b* which we shall now denote as state (κa) to indicate that it is comprised of both elastic, that is (κ) and plastic, that is (a) states. The plastic distortion tensor for this state may be written

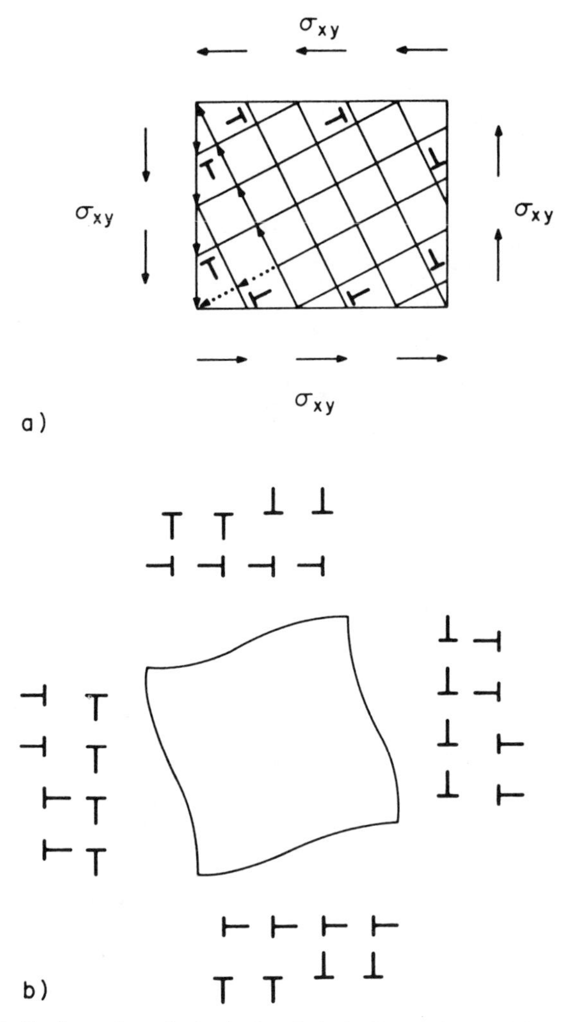

Figure 5.28 *a*) Configuration obtained after hole in Figure 5.26 is filled with stress-free matter followed by addition of a reverse external double shear stress. *b*) More complete description of *a*) in terms of both primary and secondary dislocation arrays. Reprinted by permission of Elsevier Sequoia from K. Jagannadham and M. J. Marcinkowski, (1978c), *Materials Science and Engineering*, in publication, Figure 6b.

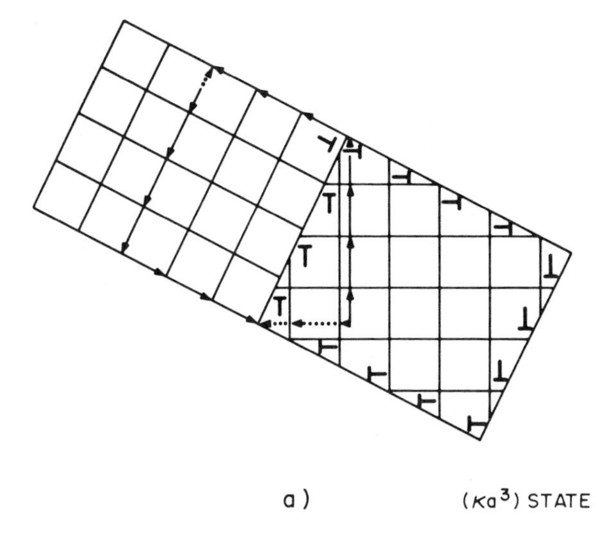

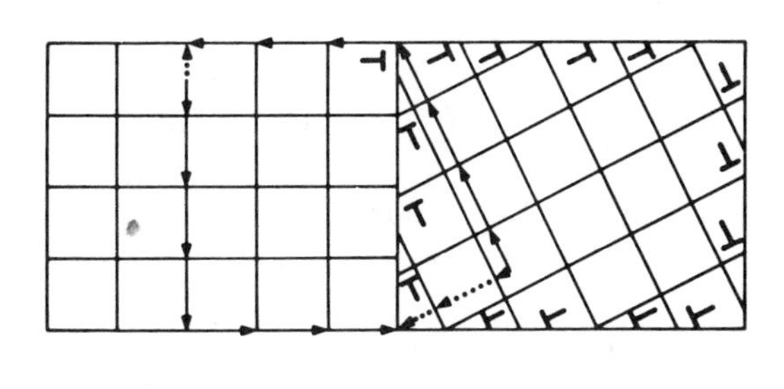

b)

Figure 5.29 *a*) Pure elastic and plastic rotations separated by an interface. *b*) Same as *a*), but after application of a reverse rotation.

as $A_K^{\kappa a}$ which has nonvanishing components

$$A_1^1 = A_3^3 = 1 \tag{5.82a}$$

and

$$A_2^2 = \{H(-x^1)\}_1 + \{2H(+x^1)\}_2 \tag{5.82b}$$

When substituted into the following expression

$$b^{\kappa a} = -\oint A_K^{\kappa a}\, dx^K \tag{5.83}$$

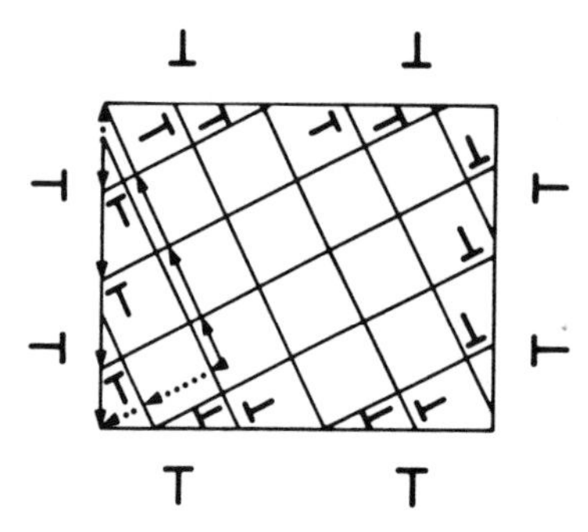

Figure 5.30 Configuration obtained after hole in figure similar to that shown in Figure 5.26 is filled with stress-free matter followed by a reverse rotation.

the distortions given by Eq. 5.82 give the dotted closure failures shown in Figure 5.16. We could also write

$$b^{\kappa a} = -\int_s S_{\kappa b\,\kappa c}^{\cdot\ \cdot\ \kappa a}\, dF^{\kappa b\,\kappa c} \tag{5.84}$$

where the only nonvanishing component of $S_{\kappa b\,\kappa c}^{\cdot\ \cdot\ \kappa a}$ is from Eqs. 4.63 and 5.82

$$S_{12}^{\cdot\ \cdot\ 2} = \left\{-\tfrac{1}{2}\delta(x^1)\right\}_1 + \left\{\left(\tfrac{1}{2}\right)2\delta(x^1)\right\}_2 = \left\{\tfrac{1}{2}\delta(x^1)\right\}_2 \tag{5.85}$$

that when substituted into Eq. 5.84 yields the same result as Eq. 5.83.

An elastic distortion $B_{\kappa a}^{K}$ can also be associated with Figure 5.16*b* whose nonvanishing components are

$$B_1^1 = B_3^3 = 1 \tag{5.86a}$$

and

$$B_2^2 = \{2H(-x^1)\}_1 + \{H(+x^1)\}_2 \tag{5.86b}$$

These distortions can in turn be used to determine the metric tensor given by

$$g_{\kappa b\,\kappa c} = B_{\kappa b}^{K} B_{\kappa c}^{K} \tag{5.87}$$

so that now both the elastic as well as the plastic strain associated with the (κa) state can be determined. It is also instructive at this point to reexamine the concept of parallel displacement. Along the lines of Eq. 5.41, we may write for the connection

$$\Gamma_{\kappa b\,\kappa c}^{\kappa a} = \left\{ {\kappa a \atop \kappa b\,\kappa c} \right\} + S_{\kappa b\,\kappa c}^{\cdot\ \cdot\ \kappa a} - S_{\kappa c\ \cdot\ \kappa b}^{\cdot\ \kappa a} + S^{\kappa a}{}_{\cdot\ \cdot\ \kappa b\,\kappa c} \tag{5.88}$$

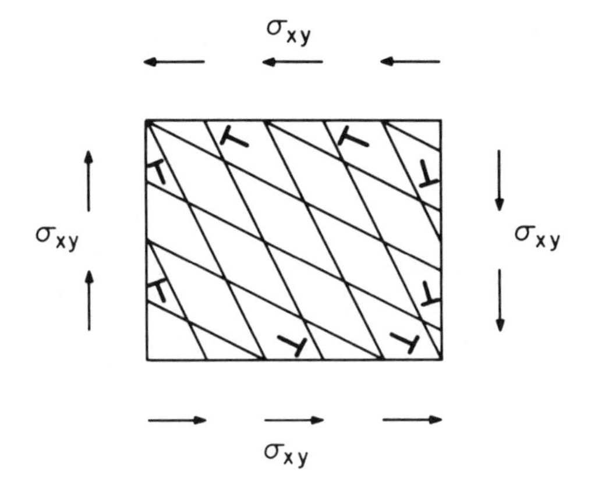

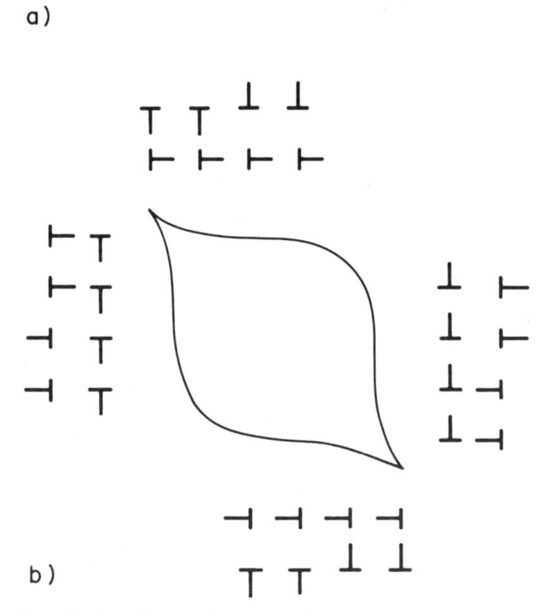

Figure 5.31 *a*) Description of pure shear of a finite body in terms of surface dislocations. *b*) More complete description of *a*) in terms of both primary and secondary dislocations. Reprinted by permission of Elsevier Sequoia from K. Jagannadham, and M. J. Marcinkowski, (1978c), *Materials Science and Engineering*, in publication, Figure 8b.

where the only possible nonvanishing components are

$$\Gamma_{12}^{2}=\left\{ {2 \atop 12} \right\}+S_{12}^{\cdot\cdot 2}-S_{2\cdot 1}^{\cdot 2}+S^{2}_{\cdot 12} \tag{5.89a}$$

and

$$\Gamma_{21}^{2}=\left\{ {2 \atop 21} \right\}+S_{21}^{\cdot\cdot 2}-S_{1\cdot 2}^{\cdot 2}+S^{2}_{\cdot 21} \tag{5.89b}$$

From Eqs. 5.86 and 5.87, the only nonvanishing components of the metric tensor are

$$g_{11}=g_{33}=1 \tag{5.90a}$$

and

$$g_{22}=\{4H(-x^{1})\}_{1}+\{H(+x^{1})\}_{2} \tag{5.90b}$$

and from the relation

$$g_{\kappa a\,\kappa b}g^{\kappa b\,\kappa c}=\delta_{\kappa a}^{\kappa c} \tag{5.91}$$

we obtain

$$g^{11}=g^{33}=1 \tag{5.92a}$$

and

$$g^{22}=\left\{\tfrac{1}{4}H(-x^{1})\right\}_{1}+\{H(+x^{1})\}_{2} \tag{5.92b}$$

From equations of the type given by Eq. 5.42, we can rewrite Eq. 5.89 as

$$\Gamma_{12}^{2}=\left\{ {2 \atop 12} \right\} \tag{5.93a}$$

and

$$\Gamma_{21}^{2}=\left\{ {2 \atop 12} \right\}+2S_{21}^{\cdot\cdot 2} \tag{5.93b}$$

The Christoffel symbol of the second kind may be written as

$$\left\{ {\kappa a \atop \kappa b\,\kappa c} \right\}=\tfrac{1}{2}g^{\kappa a\,\kappa d}(\partial_{\kappa b}g_{\kappa c\,\kappa d}+\partial_{\kappa c}g_{\kappa b\,\kappa d}-\partial_{\kappa d}g_{\kappa b\,\kappa c}) \tag{5.94}$$

where the only possible nonvanishing components are

$$\left\{ {2 \atop 12} \right\}=\left\{ {2 \atop 21} \right\}=\tfrac{1}{2}g^{2\kappa d}(\partial_{1}\lambda_{2\kappa d}+\partial_{2}g_{1\kappa d}-\partial_{\kappa d}g_{12}) \tag{5.95a}$$

which in view of Eqs. 5.90 and 5.92, further reduces to

$$\left\{ {2 \atop 12} \right\} = \left\{ {2 \atop 21} \right\} = \tfrac{1}{2} g^{22}(\partial_1 g_{22}) \tag{5.95b}$$

or

$$\left\{ {2 \atop 12} \right\} = \left\{ {2 \atop 21} \right\} = \left\{ -\tfrac{1}{2} H(-x^1)\delta(x^1) \right\}_1 + \left\{ \tfrac{1}{2} H(+x^1)\delta(x^1) \right\}_2 = 0 \tag{5.95c}$$

The above result could have also been obtained using the concept of parallel displacement, that is,

$$dC^{\kappa a} = -\Gamma^{\kappa a}_{\kappa b\,\kappa c} C^{\kappa b}\, dx^{\kappa c} \tag{5.96}$$

where it is obvious that

$$dC^2 = -\Gamma^2_{12} C^1\, dx^2 = 0 \tag{5.97}$$

so that $\Gamma^2_{12} = 0$, and from Eq. 5.93a,

$$\left\{ {2 \atop 12} \right\} = \left\{ {2 \atop 21} \right\} = 0 \tag{5.98}$$

so that from Eq. 5.85

$$\Gamma^2_{21} = 2S_{21}^{\cdot\cdot 2} = \{\delta(x^1)\}_2 \tag{5.99}$$

Thus, in terms of the following relation;

$$dC^2 = -\Gamma^2_{21} C^2\, dx^1 \tag{5.100}$$

when a vector of strength $C^2 = 4$ in region 1 of Figure 5.16*b* is moved along x^1 across the boundary, it changes to $C^2 = 8$ in region 2, so that $\Delta C^2 = +4$. It is important to note that the coordinates associated with $C^{\kappa b}$ must be those of the individual regions.

An analysis of the (κa^1) state shown in Figure 5.21*b* follows closely along the lines as that for the (κa) state. It is, however, instructive to consider the (κa^2) state shown in Figure 5.24*a*. In this case the nonvanishing components of the distortion tensor $A_K^{\kappa a^2}$ become

$$A_1^1 = A_2^2 = A_3^3 = 1 \tag{5.101a}$$

and

$$A_2^1 = \{0\}_1 + \{H(x^1)\tan\theta\}_2 \tag{5.101b}$$

that when substituted into

$$b^{\kappa a^2} = -\oint A_K^{\kappa a^2} dx^K \tag{5.102}$$

yields the closure failure shown dotted in Figure 5.24*a*. Equation 5.102 could also be rewritten as

$$b^{\kappa a^2} = -\int_s S_{\kappa b^2 \kappa c^2}^{\cdot\ \cdot\ \kappa a^2} dF^{\kappa b^2 \kappa c^2} \tag{5.103}$$

where the only nonvanishing component of the torsion tensor is

$$S_{12}^{\cdot\cdot 1} = \{0\}_1 + \left\{\tfrac{1}{2}\delta(x^1)\tan\theta\right\}_2 \tag{5.104}$$

and when substituted into Eq. 5.103 yields the same result as Eq. 5.102.

The elastic distortion tensor $B_{\kappa a^2}^{K}$ associated with the (κa^2) state has nonvanishing components

$$B_1^1 = B_2^2 = B_3^3 = 1 \tag{5.105a}$$

and

$$B_2^1 = \{H(x^1)\tan\theta\}_1 + \{0\}_2 \tag{5.105b}$$

It is now easy to show, employing an expression of the type given by Eq. 5.88, that the only possible nonvanishing components of the connection are

$$\Gamma_{12}^1 = \left\{ {1 \atop 12} \right\} + S_{12}^{\cdot\cdot 1} - S_{2\cdot 1}^{\cdot 1} + S^1_{\cdot 12} \tag{5.106a}$$

and

$$\Gamma_{21}^1 = \left\{ {1 \atop 21} \right\} + S_{21}^{\cdot\cdot 1} - S_{1\cdot 2}^{\cdot 1} + S^1_{\cdot 21} \tag{5.106b}$$

Since the metric tensor can be written as

$$g_{\kappa a^2 \kappa b^2} = \left\{\delta_{\kappa a^2}^{\kappa^2}\delta_{\kappa b^2}^{\kappa^2} g_{\kappa^2\lambda^2} H(-x^1)\right\}_1 + \left\{\delta_{\kappa a^2 \kappa b^2} H(x^1)\right\}_2 \tag{5.107}$$

where $g_{\kappa^2\lambda^2}$ in the above expression is given by Eq. 2.25, it is a rather straightforward matter to show that

$$\Gamma_{12}^1 = 2S_{12}^{\cdot\cdot 1} \tag{5.108a}$$

and

$$\Gamma^1_{21} = 0 \tag{5.108b}$$

so that from Eq. 5.104 we can write

$$\Gamma^1_{12} = \{0\}_1 + \{\delta(x^1)\tan\theta\}_2 \tag{5.109}$$

so that from the following equation for parallel displacement

$$dC^1 = \Gamma^1_{12} C^2 dx^1 \tag{5.110}$$

we can say that when a vector $C^2 = 4$ in region 1 of Figure 5.24*a* is moved along x^1 across the interface, it undergoes a change $\Delta C^1 = 2$ in region 2. This change is in fact the dotted line shown in the figure. Similar analyses follow for the (κa^4) and (κa^3) states shown in Figure 5.27*a* and 5.29*a*, respectively.

REVIEW

The concept of a surface dislocation is by far the most important development introduced in the present text and forms the backbone of the present theory dealing with the mechanical behavior of matter. It is most easily understood by referring to Figure 5.11*a* that shows a single quantized edge type crystal lattice dislocation contained within the interior of a crystal. Such a dislocation gives rise to long-range internal stresses that extend all the way to the surface of the crystal. This is reflected schematically in Figure 5.11*a* by the bending of the vertical lines that takes place between points 2″–1′ and 3″–4′. The bending of these lines in turn means that the surfaces of the body are not stress free and thus violates the condition that the surface in general must be stress free. However, this stress free boundary requirement can be met by the introduction of an array of surface dislocation on the faces of the body as illustrated in Figure 5.11*b*. These surface dislocations are somewhat more complex than the interior crystal lattice dislocation, as is shown by the single dislocation in the inset toward the right of Figure 5.11*b*. In particular, this surface dislocation may be viewed as a dislocation dipole. One of the edge type dislocations comprising the dipole is shown dotted and may be viewed as possessing no stress field, but instead gives rise to the ledge shown by the dotted arrow. The second dislocation comprising the dipole is shown by solid lines and has associated with it the normal stress field of an edge type dislocation lying within a crystal; however, it has no free surface associated with it. Another way of looking at these surface dislocations or dipoles is in terms of a link or transition zone between the distorted interior of the body and the undistorted surface. In particular, they allow the outer surface fibers of the body to remain vertical. Another important feature of the surface dislocations is that they are not quantized but are continuously distributed and have infinitesimally small Burger vectors. We shall see in Chapter 11 that such dislocation arrays are most easily treated in terms of the theory of continuously distributed dislocations.

Another important property of the surface dislocations is that their total displacement or Burgers vector, as represented by the solidly drawn dislocations of the dipole, must be balanced by that of the internal crystal lattice dislocation. This statement therefore corresponds to a very powerful conservation law embodied in Eq. 5.80 that states that the sum of the Burgers vectors of the internal and surface dislocations must add up to zero. A very important consequence of this law is best seen by considering the movement of the crystal lattice dislocation to the surface of the body. Under those conditions, the surface dislocations gradually attract the crystal lattice dislocation. Finally, when the crystal dislocation meets the surface, the surface dislocations combine with it in the manner shown in Figure 5.11*d* to produce a single stress-free ledge of the same Burgers vector as that possessed by the quantized crystal lattice dislocation. At this point, is to be emphasized that combination of the crystal lattice dislocation with the surface dislocations does not mean that the crystal lattice dislocation has left the crystal. On the contrary, as Figure 5.11*d* shows, it is still present as shown by the dotted edge dislocation symbol, but in modified form, that is, it has completely lost its stress field. In the language of differential geometry, one may say that the crystal lattice dislocation in Figure 5.11 has associated with it a torsion tensor $S_{lm}^{\cdot\cdot k}$, but no anholonomic object $\Omega_{lm}^{\cdot\cdot k}$, that is, no free surface, since the anholonomic object is a measure of the amount of free surface. On the other hand, the two dislocations comprising the surface dislocation dipole each have associated with them a torsion tensor, whereas only the one shown dotted has in addition an anholonomic object, since it is only this dislocation that has associated with it a free surface. Upon combination of the crystal lattice dislocation with the surface dislocation dipoles to yield Figure 5.11*d*, one may view the torsion tensor associated with the solid dislocation symbols in Figure 5.11*b* as just canceling with each other leaving only the dotted one shown in Figure 5.11*d* that now possesses both torsion as well as an anholonomic object such that $S_{cb}^{\cdot\cdot a}=\Omega_{cb}^{\cdot\cdot a}$ that is in fact Eq. 5.81. This means, in view of a surface integral of the type given by Eq. 5.44, the Burgers vector associated with $S_{cb}^{\cdot\cdot a}$ just balances that associated with the newly created free surface, that obtained from $\Omega_{cb}^{\cdot\cdot a}$.

The surface dislocation array shown in Figure 5.11*b*, although it nearly satisfies the stress free boundary conditions, does not do so exactly. In order to accomplish this, it is necessary to add a secondary array of surface dislocations with Burgers vectors normal to those of the primary array such as shown in Figure 5.14*b*. In this case the dislocations occur in pairs of equal but opposite signs. For simplicity, Figure 5.14*b* shows only that part of the dipole which actually gives rise to the stress field, that is the one drawn solid.

It is now possible to use the concept of a surface dislocation to satisfy the stress-free boundary conditions associated with any internally stressed body since, as Chapter 7 shows, any internal stress can be described in terms of some suitable continuously distributed array of dislocations of infinitesimal strength.

The power of the surface dislocation method is next demonstrated by using them to solve problems associated with various states of external stress. Consider for example the body subjected to a uniform tensile stress σ_{yy} in Figure 5.19*a*. A portion of the body can next be removed as illustrated in Figure 5.19*b* so as to create the hole shown at the left of this figure. In order to satisfy the boundary conditions a set of surface dislocation dipoles must be placed on both the surface of the hole as well as the cut out body. If, however, extra matter is added to the finite body, as shown toward the right of Figure 5.19*c*, this body can be made stress free and placed back into the hole, as shown in Figure 5.19*c*. An equal but opposite uniform compressive stress σ_{yy} can now be applied to the reassembled body, as illustrated in Figure 5.19*e*. If the finite body is again cut out of the infinite body, the configuration shown toward the right of Figure 5.19*f* obtains. In particular, the state of stress within the interior of a rectangular body subjected to a uniform compressive stress on its top and bottom faces can be represented in terms of an array of surface dislocations on all its four faces. For simplicity, only one such surface dislocation is shown in Figure 5.19*f*. In a similar manner, the state of stress within a stressed body containing a hole, such as shown in Figure 5.19*b*, can also be described in terms of some suitable array of surface dislocations placed on the surface of the hole. We are thus led to a very powerful result that the stress-free boundary conditions associated with any given state of stress can be exactly described in terms of a suitable continuous distribution of dislocations. In general, for arbitrarily shaped bodies, the solutions are difficult if not impossible to carry out analytically. However, by allowing the surface dislocations to possess discrete Burgers vectors and be finite in number, such problems can be solved quite easily by means of computer techniques.

In concluding this section, it is instructive to reflect on the fundamental concept by which dislocations are able to be used in the solution of boundary value problems. The reason for this is that edge and screw type dislocations can generate various components of stress, and combinations of both can generate all six components of stress. Thus, any suitable distribution of such dislocations can be used to describe any given state of stress. Another way of looking at this problem is in terms of finding the Airy stress functions or Green's function associated with a single dislocation. However, this has already been done once and for all. Now rather than find such functions for more complex problems, which is by the way extremely difficult, we need now only find a suitable dislocation distribution. In most cases the dislocation distribution can be guessed at from physical considerations alone. It is next only a matter of finding the positions at which the total stress on each dislocation in the array is zero that can in turn be done numerically. A great mystery to the present author is why this simple but powerful concept has gone unnoticed for so long.

Finally, the concept of a surface dislocation enables the purely plastic states treated in Chapter 3 to be understood more easily. In particular, all of the plastic states may be generated by allowing crystal lattice dislocations within the body, such as those considered in Chapter 4, to move to the surface of the crystal and thus combine with their respective surface dislocations. This combination is equivalent to compensating the closure failure associated with the crystal lattice dislocation by creating an amount of new surface whose corresponding closure failure just matches that of the dislocation. In this way, all of the pure plastic states considered in Chapter 3 may be viewed in terms of dislocations. Up till now, it was originally thought that the crystal lattice dislocations, upon reaching the surface of the crystal, effectively left the body. As stated earlier, they in fact remain on the surface of the body and account for the change in shape of the external surface of the body. This is the only way that dislocations can be detected since the interior of the body is maintained perfect as easily seen by reference to Figure 5.2*a*.

Surface dislocations are also of use in describing the distortions associated with internal surfaces such as two-phase interfaces and grain boundaries as Chapters 6 and 13 show. Equally exciting, is that finding that for the first time, a complete theory of surface tension for liquids as well as solids is possible. In this case it is the surface dislocations that are directly responsible for the surface tension of a finite body. This will be treated in detail in Chapter 13. We thus have the foundations for formulating a new and powerful method of treating the behavior of liquids.

CHAPTER 6

Grain Boundaries and Two-Phase Interfaces

In view of the success of the free surface dislocations introduced for the first time in the previous chapter, it seems quite logical to extend these concepts to internal surfaces. The ones that immediately come to mind are grain boundaries and two-phase interfaces. It is noted in what follows that internal boundaries, for the most part, have elastic strains associated with them. These strains can be released by separating or tearing the two grains, that comprise the boundary, from one another. Most theories to date dealing with such boundaries in effect apply to torn boundaries that are also synonymous with boundaries formed by rigid operations. Such theories can therefore say nothing of the strain energy of these internal surfaces.

Closely related to the surface dislocations treated in the previous section are those that separate one phase from another or one grain from another. One such state is shown in Figure 6.1*a* and is in fact a symmetric tilt type grain boundary (Marcinkowski, 1977b). It is denoted as state (k^{I1}) where the kernel k is used to denote a halonomic dislocated state, while the superscript I serves to identify the state as belonging to the interface class, that is, interphase boundaries, grain boundaries, and so forth. The (k^{I1}) state may be generated from the (K) state shown in Figure 5.16*a* by means of the distortion tensor $A_K^{k^{I1}}$ that has nonvanishing components

$$A_1^1 = A_2^2 = A_3^3 = 1 \qquad (6.1a)$$

and

$$A_2^1 = \{H(-x^1)\tan\theta\}_1 + \{-H(+x^1)\tan\theta\}_2 \qquad (6.1b)$$

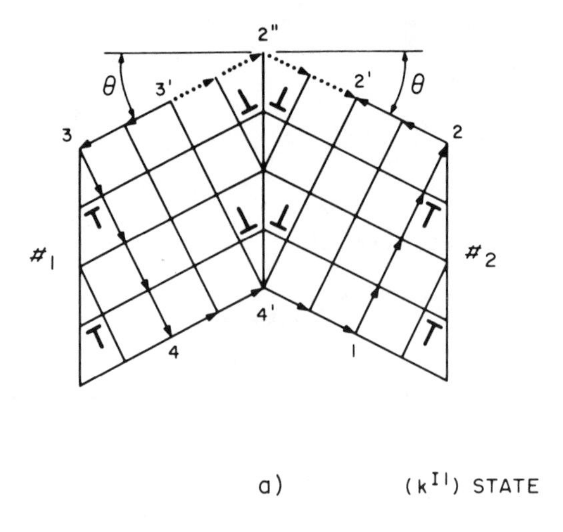

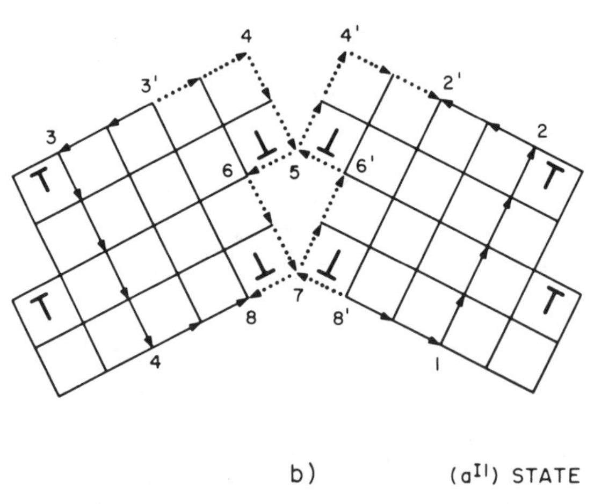

Figure 6.1 Symmetric tilt boundary. *a*) With elastic strain present *b*) With elastic strain removed.

where again the curly brackets are used to denote the fact that both grains can be considered separately. The dislocation content of the grain boundary in Figure 6.1*a* may be obtained from the following line integral:

$$b^{k'1} = -\oint A_K^{k'1}\, dx^K \tag{6.2}$$

which has the nonvanishing component given by

$$b^1 = -\int_3^4 A_2^1\,dx^2 - \int_1^2 A_2^1\,dx^2 \tag{6.3a}$$

which in view of Eqs. 6.1 becomes

$$b^1 = \{4\tan\theta\}_1 + \{4\tan\theta\}_2 = \{2\}_1 + \{2\}_2 \tag{6.3b}$$

or in terms of Figure 6.1*a*

$$b^1 = \underset{3'-2''}{\Delta x^1} + \underset{2''-2'}{\Delta x^1} \tag{6.3c}$$

The first term in the last equation corresponds to the dotted closure failure 3′–2″ within grain #1 while the second term corresponds to the dotted closure failure 2″–2′ in grain #2.

We could also write the grain boundary closure failure in terms of the following surface integral:

$$b^{k^{I1}} = -\int_s S_{l^{I1}m^{I1}}^{\cdot\;\cdot\;k^{I1}}\,dF^{l^{I1}m^{I1}} \tag{6.4}$$

where from an expression of the type given by Eq. 4.91, the only nonvanishing component of $S_{l^{I1}m^{I1}}^{\cdot\;\cdot\;k^{I1}}$ becomes

$$S_{12}^{\cdot\,\cdot\,1} = \left\{-\left(\tfrac{1}{2}\right)\delta(x^1)\tan\theta\right\}_1 + \left\{-\left(\tfrac{1}{2}\right)\delta(x^1)\tan\theta\right\}_2 \tag{6.5}$$

which when substituted into Eq. 6.4 yields the same result as that given by Eq. 6.3. It should be emphasized that the closure failure given by Eq. 6.3*b* is expressed in component form. When expressed in vector form it becomes

$$\mathbf{b} = \{2\mathbf{e}_1\}_1 + \{2\mathbf{e}_1\}_2 \tag{6.6}$$

where $\mathbf{e}_{k^{I1}}$ are the base vectors associated with the (k^{I1}) state.

The torsion tensor given by Eq. 6.5 can also be used to determine the grain boundary dislocation density by means of the following expression:

$$\alpha^{n^{I1}k^{I1}} = -\varepsilon^{n^{I1}l^{I1}m^{I1}} S_{l^{I1}m^{I1}}^{\cdot\;\cdot\;k^{I1}} \tag{6.7}$$

and more specifically

$$\alpha^{31} = -2S_{12}^{\cdot\,\cdot\,1} = \{\delta(x^1)\tan\theta\}_1 + \{\delta(x^1)\tan\theta\}_2 \tag{6.8}$$

It is apparent that the above dislocation density, because it is in the form of a Dirac delta function, vanishes everywhere except at $x^1=0$. It is also apparent from Eq. 4.115 that α^{31} is given in terms of unit area. However, when Eq. 6.8 is integrated with respect to x^1 we obtain

$$\alpha^{31}=\{\tan\theta\}_1+\{\tan\theta\}_2=\left\{\tfrac{1}{2}\right\}_1+\left\{\tfrac{1}{2}\right\}_2 \tag{6.9}$$

so that now α^{31} is expressed in terms of a unit length. In terms of Figure 6.1*a*, Eq. 6.9 may be written as

$$\alpha^{31}=\left\{\frac{\underset{3'-2''}{\Delta x^1}}{\underset{4'-3'}{\Delta x^2}}\right\}_1+\left\{\frac{\underset{2''-2'}{\Delta x^1}}{\underset{4'-2'}{\Delta x^2}}\right\}_2 \tag{6.10}$$

The elastic distortions associated with the (k^{I1}) state of Figure 6.1*a* can be relaxed so as to generate its anholonomic counterpart which is shown in Figure 6.1*b* as state (a^{I1}). Kernel *a* again is used to denote anholonomic coordinates. It is apparent that the (a^{I1}) state has now associated with it free surfaces at the grain boundary. The nonvanishing components of the distortion tensor $A_K^{a^{I1}}$ now become

$$A_3^3=1 \tag{6.11a}$$

$$A_1^1=A_2^2=\{H(-x^1)\}_1+\{H(+x^1)\}_2 \tag{6.11b}$$

$$A_2^1=\{H(-x^1)\tan\theta\}_1+\{H(+x^1)\tan\theta\}_2 \tag{6.11c}$$

When used in conjunction with the following equations

$$b^{a^{I1}}=-\oint A_K^{a^{I1}}dx^K \tag{6.12a}$$

$$\underset{s}{b}^{a^{I1}}=\oint A_K^{a^{I1}}dx^K \tag{6.12b}$$

Eq. 6.12a yields the same result as that given by Eqs. 6.3. On the other hand, Eq. 6.12b, when used with Eqs. 6.11b and 6.11c, gives

$$\underset{s}{b}^1=\{-2\}_1+\{-2\}_2=\underset{5-6}{\Delta x^1}+\underset{6'-5}{\Delta x^1} \tag{6.13a}$$

and

$$\underset{s}{b}^2=\{-4\}_1+\{4\}_2=\left\{\underset{4-5}{\Delta x^2}+\underset{6-7}{\Delta x^2}\right\}_1+\left\{\underset{7-6'}{\Delta x^2}+\underset{5-4'}{\Delta x^2}\right\}_2 \tag{6.13b}$$

Both of these last results correspond to the surface closure failures shown in Figure 6.1*b*.

The results given by Eq. 6.12 could be written more compactly in terms of the following surface integral:

$$b^{a''} = -\int_S \left(S_{c''b''}^{\cdot\cdot\ a''} - \Omega_{c''b''}^{\cdot\cdot\ a''}\right) dF^{c''b''} \tag{6.14}$$

We can also write

$$S_{12}^{\cdot\cdot\ 1} = \Omega_{12}^{\cdot\cdot\ 1} \tag{6.15}$$

where $S_{12}^{\cdot\cdot\ 1}$ is the same as that given by Eq. 6.5, while from an expression of the type given by Eq. 5.66, we obtain

$$\Omega_{12}^{\cdot\cdot\ 2} = \left\{-\left(\tfrac{1}{2}\right)\delta(x^1)\right\}_1 + \left\{\left(\tfrac{1}{2}\right)\delta(x^1)\right\}_2 \tag{6.16}$$

Equation 6.15, when substituted into Eq. 6.14, yields the same results as those given by Eqs. 6.3 and 6.13a, while Eq. 6.16 when substituted into Eq. 6.14, yields the same result as that given by Eq. 6.13b. It is also clear that we can write an expression for the grain boundary dislocation density $\alpha^{d''a''}$ similar to that given by Eq. 6.7. In addition, we can also write a free surface density given by

$$\beta^{d''a''} = \varepsilon^{d''c''b''} \Omega_{c''b''}^{\cdot\cdot\ a''} \tag{6.17}$$

which in view of Eqs. 6.15 and 6.16, yields

$$\beta^{31} = \{-\tan\theta\}_1 + \{-\tan\theta\}_2 \tag{6.18a}$$

and

$$\beta^{32} = \{-1\}_1 + \{1\}_2 \tag{6.18b}$$

The last two equations have been integrated with respect to x^1 so that they represent surface densities. In terms of Figure 6.1*b* we may write

$$\beta^{31} = \left\{\frac{\underset{5-6}{\Delta x^1} + \underset{7-8}{\Delta x^1}}{\underset{8-3'}{\Delta x^2}}\right\}_1 + \left\{\frac{\underset{6'-5}{\Delta x^1} + \underset{8'-7}{\Delta x^1}}{\underset{8'-2'}{\Delta x^2}}\right\}_2 \tag{6.19a}$$

and

$$\beta^{32} = \left\{ \frac{\underset{4-5}{\Delta x^2} + \underset{6-7}{\Delta x^2}}{\underset{8-3'}{\Delta x^2}} \right\}_1 + \left\{ \frac{\underset{5-4'}{\Delta x^2} + \underset{7-6'}{\Delta x^2}}{\underset{8'-2'}{\Delta x^2}} \right\}_2 \qquad (6.19b)$$

If the (K) state of Figure 5.16a were simply torn or broken along the dotted line and symmetrically rotated, we would obtain the (a^{I2}) state shown in Figure 6.1c. In this case the torsion tensor $S_{c^{I2}b^{I2}}^{\cdot\;\cdot\;a^{I2}}$ would be zero

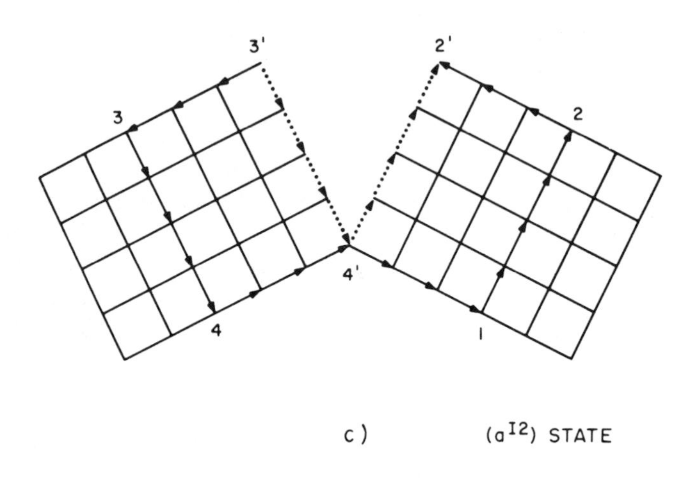

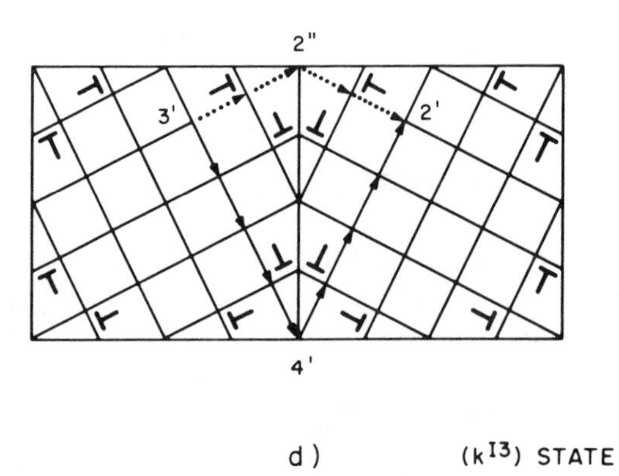

Figure 6.1 (Continued) c) Symmetrically torn body. d) Symmetric tilt boundary generated by double shear.

since no dislocations are present. However, $\Omega_{c^{l2}b^{l2}}^{\cdot\cdot a^{l2}}$ possesses a nonvanishing component $\Omega_{12}^{\cdot\cdot 2}$ that is identical to that given by Eq. 6.16, and that when used in the following equation;

$$\underset{s}{b}^{a^{l2}} = \int_s \Omega_{c^{l2}b^{l2}}^{\cdot\cdot a^{l2}} dF^{c^{l2}b^{l2}} \tag{6.20}$$

yields the closure failures shown by the dotted lines in Figure 6.1*c*.

Consider the (k^{l3}) state shown in Figure 6.1*d* that may be generated from the (K) state of Figure 5.16*a* by means of the following distortion tensor:

$$A_K^{k^{l3}} = \left\{ H(-x^1)\underset{1}{A_K^{k^{l3}}} \right\}_1 + \left\{ H(+x^1)\underset{2}{A_K^{k^{l3}}} \right\}_2 \tag{6.21}$$

where

$$\underset{1}{A_K^{k^{l3}}} = \delta_L^{k^{l3}} \delta_K^{\kappa^4} B_{\kappa^4}^L \tag{6.22a}$$

and where $B_{\kappa^4}^L$ is given by Eq. 2.30, while

$$\underset{2}{A_K^{k^{l3}}} = \underset{1}{A_{k^{l3}}^K} \tag{6.22b}$$

These distortions can be combined with the following line integral:

$$b^{k^{l3}} = -\oint A_K^{k^{l3}} dx^K \tag{6.23}$$

to obtain the grain boundary dislocation content whose Burgers vectors are shown by the dotted lines in Figure 6.1*d*. In this particular case, the reference circuit 1–2–3–4 in Figure 5.16*a* has been shrunk to the position of the boundary. It is clear that the grain boundary in Figure 6.1*d* is identical to that shown in Figure 6.1*a* and arises from the fact that the component A_1^2 in Eq. 6.21 does not enter into the calculation. If, however, the distortion given by Eq. 6.21 were to take place such that grain #1 were entirely enclosed by grain #2, as shown in Figure 6.2, then the component A_1^2 would be required to determine the dislocation content with respect to the horizontal portions of the grain boundary such as shown by the dotted arrows in this figure. In this case, x^1 in Eq. 6.21 would also have to be replaced by x^2. It also follows that the torsion tensor and dislocation density associated with the grain boundary in Figure 6.1*d* are also identical to those determined for the grain boundary in Figure 6.1*a*.

Next let us consider the symmetric tilt boundary shown in Figure 6.3 and labeled as state (k^{I4}). It may be viewed as generated from the (k^{I3}) state of Figure 6.1d as follows

$$A_K^{k^{I4}} = A_K^{k^{I3}} A_{k^{I3}}^{k^{I4}} \tag{6.24}$$

where

$$A_{k^{I3}}^{k^{I4}} = \begin{bmatrix} \cos\theta & 0 & 0 \\ 0 & \cos\theta & 0 \\ 0 & 0 & 1 \end{bmatrix} \tag{6.25}$$

It is apparent that when the above relation is applied to Eq. 6.21 we obtain

$$\underset{1}{A_K^{k^{I4}}} = \begin{bmatrix} \cos\theta & -\sin\theta & 0 \\ \sin\theta & \cos\theta & 0 \\ 0 & 0 & 1 \end{bmatrix} \tag{6.26}$$

with a similar relation for $\underset{2}{A_K^{k^{I4}}}$ which differs from the above equation by an interchange of sign in the $\sin\theta$ components. It thus follows that the grain boundary in Figure 6.3 is generated from the (K) state of Figure 5.16a by a pair of pure plastic rotations of the type described in Section 3. In order to determine the dislocation content associated with such a

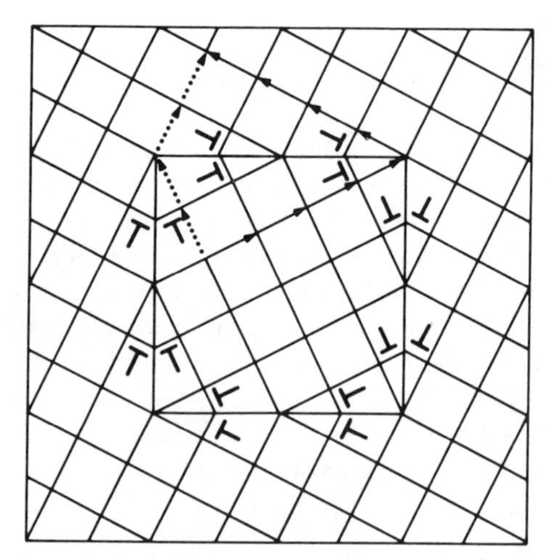

Figure 6.2 Symmetric grain boundary associated with a grain entirely enclosed within a crystal.

boundary, we may write

$$b^{k^{I4}} = -\oint A_K^{k^{I4}} dx^K \tag{6.27}$$

which together with Eq. 6.24 yields

$$b^1 = -\int_3^4 A_2^1 dx^2 - \int_1^2 A_2^1 dx^2 \tag{6.28a}$$

or

$$b^1 = \{4\sin\theta\}_1 + \{4\sin\theta\}_2 \tag{6.28b}$$

or in terms of Figure 6.3

$$b^1 = \left\{ \underset{3'-2''}{\Delta x^1} \right\}_1 + \left\{ \underset{2''-2'}{\Delta x^1} \right\}_2 \tag{6.28c}$$

In terms of a surface integral, Eq. 6.27 becomes

$$b^{k^{I4}} = -\int_s S_{l^{I4}m^{I4}}^{\cdot\cdot\ k^{I4}} dF^{l^{I4}m^{I4}} \tag{6.29}$$

Since the only nonvanishing component of $S_{l^{I4}m^{I4}}^{\cdot\cdot\ k^{I4}}$ is

$$S_{12}^{\cdot\cdot 1} = \left\{ -\left(\tfrac{1}{2}\right)\delta(x^1)\sin\theta \right\}_1 + \left\{ -\left(\tfrac{1}{2}\right)\delta(x^1)\sin\theta \right\}_2 \tag{6.30}$$

Eq. 6.29 yields the same result as that given by Eq. 6.28. The grain boundary dislocation density can next be written as

$$\alpha^{n^{I4}k^{I4}} = -\varepsilon^{n^{I4}l^{I4}m^{I4}} S_{l^{I4}m^{I4}}^{\cdot\cdot\ k^{I4}} \tag{6.31}$$

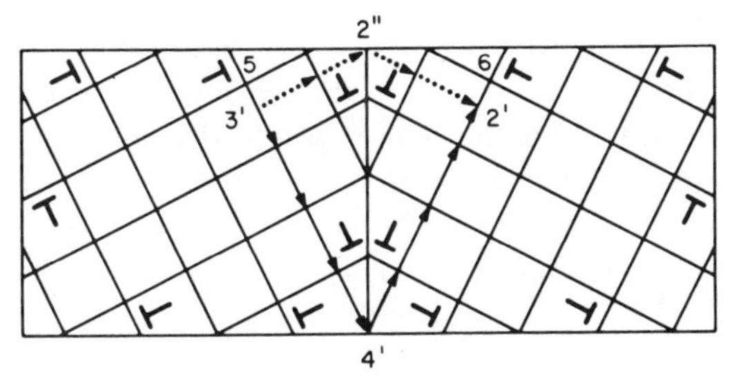

Figure 6.3 Symmetric tilt boundary generated by pure plastic rotation.

which in view of Eq. 6.30 yields

$$\alpha^{31}=\{\sin\theta\}_1+\{\sin\theta\}_2 \tag{6.32}$$

In terms of Figure 6.3 the above equation may be written as

$$\alpha^{31}=\left\{\frac{\underset{3'-2''}{\Delta x^1}}{\underset{4'-5}{\Delta x^2}}\right\}_1+\left\{\frac{\underset{2''-2'}{\Delta x^1}}{\underset{4'-6}{\Delta x^2}}\right\}_2 \tag{6.33a}$$

However, it is incorrect to write

$$\alpha^{31}=\left\{\frac{\underset{3'-2''}{\Delta x^1}}{\underset{4'-2''}{\Delta x^2}}\right\}_1+\left\{\frac{\underset{2''-2'}{\Delta x^1}}{\underset{4'-2''}{\Delta x^2}}\right\}_2 \tag{6.33b}$$

Again, as in Figure 6.1d, the grain boundary shown in Figure 6.3 has associated with it elastic strains.

Figure 6.4 shows a symmetric tilt boundary that was generated from the (K) state by means of the distortion tensor $A_K^{k^{I5}}$ with nonvanishing components

$$A_1^1=A_2^2=A_3^3=1 \tag{6.34}$$

and

$$A_2^1=\left\{-H(-x^1)\tan\left(\tfrac{\pi}{2}-\theta\right)\right\}_1+\left\{H(+x^1)\tan\left(\tfrac{\pi}{2}-\theta\right)\right\}_2 \tag{6.35a}$$

or alternately as

$$A_2^1=\{-H(-x^1)\cot\theta\}_1+\{H(+x^1)\cot\theta\}_2 \tag{6.35b}$$

In spite of the fact that the atom configuration associated with the grain boundary in Figure 6.4 is identical to that in Figure 6.1a, the dislocation contents are markedly different. More specifically, when Eq. 6.35 is substituted into the following relation:

$$b^{k^{I5}}=-\oint A_K^{k^{I5}}\,dx^K \tag{6.36}$$

we obtain

$$b^1=\{-4\cot\theta\}_1+\{-4\cot\theta\}_2=\{-8\}_1+\{-8\}_2 \tag{6.37a}$$

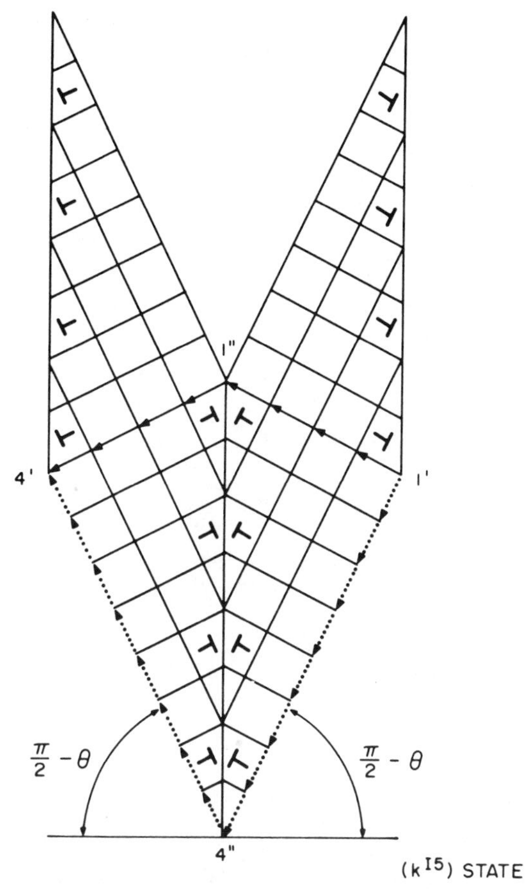

Figure 6.4 Symmetric tilt boundary generated by simple shear.

or in terms of Figure 6.4

$$b^1 = \left\{ \underset{4''-4'}{\Delta x^1} \right\}_1 + \left\{ \underset{1'-4''}{\Delta x^1} \right\}_2 \tag{6.37b}$$

In terms of a surface integral, Eq. 6.36 becomes

$$b^{k^{I5}} = -\int_s S_{l^{I5}m^{I5}}^{\cdot\;\cdot\;k^{I5}}\, dF^{l^{I5}m^{I5}} \tag{6.38}$$

where the only nonvanishing component of the torsion tensor is

$$S_{12}^{\cdot\cdot 1} = \left\{ \left(\tfrac{1}{2}\right)\delta(x^1)\cot\theta \right\}_1 + \left\{ \left(\tfrac{1}{2}\right)\delta(x^1)\cot\theta \right\}_2 \tag{6.39}$$

which when substituted into Eq. 6.38 yields the same result as that given by Eq. 6.37. The dislocation density associated with the grain boundary of Figure 6.4 can also be written as

$$\alpha^{n^{I5}k^{I5}} = -\varepsilon^{n^{I5}l^{I5}m^{I5}} S_{l^{I5}m^{I5}}^{\cdot\ \cdot\ k^{I5}} \tag{6.40}$$

which in view of Eq. 6.39, yields

$$\alpha^{31} = \{-\cot\theta\}_1 + \{-\cot\theta\}_2 = \{-2\}_1 + \{-2\}_2 \tag{6.41}$$

or in terms of Figure 6.4

$$\alpha^{31} = \left\{ \frac{\underset{4''-4}{\Delta x^1}}{\underset{4'-1''}{\Delta x^2}} \right\}_1 + \left\{ \frac{\underset{1'-4''}{\Delta x^1}}{\underset{1'-1''}{\Delta x^2}} \right\}_2 \tag{6.42}$$

A grain boundary similar to that shown in Figure 6.4 can also be generated by a double shear and is designated as state (k^{I6}) in Figure 6.5. Such a boundary can be produced via the following distortion tensor:

$$A_K^{k^{I6}} = \left\{ H(-x^1) \underset{1}{A_K^{k^{I6}}} \right\}_1 + \left\{ H(+x^1) \underset{2}{A_K^{k^{I6}}} \right\}_2 \tag{6.43}$$

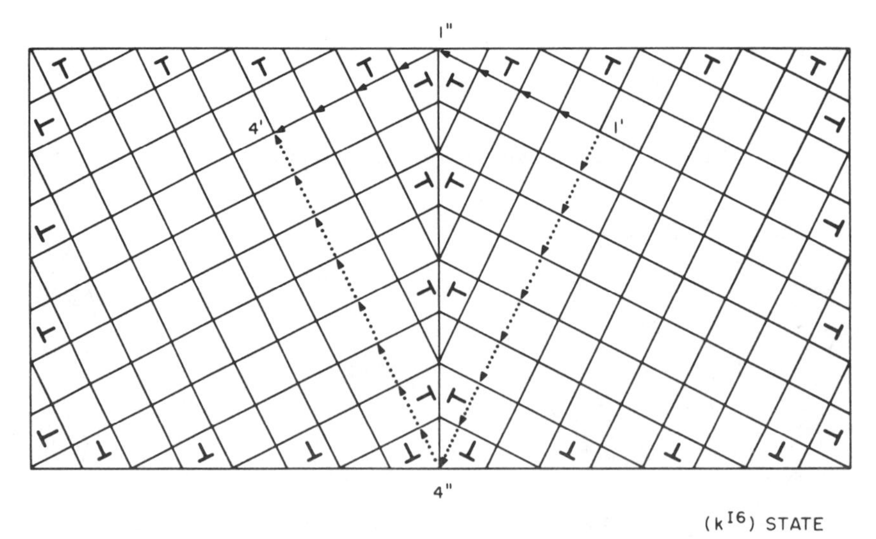

Figure 6.5 Symmetric tilt boundary similar to that shown in Figure 6.4 but generated by a double shear.

where

$$\underset{1}{A}_K^{k^{I6}} = \begin{bmatrix} 1 & \cot\theta & 0 \\ -\cot\theta & 1 & 0 \\ 0 & 0 & 1 \end{bmatrix} \tag{6.44a}$$

and

$$\underset{2}{A}_K^{k^{I6}} = \underset{1}{A}_{k^{I6}}^K \tag{6.44b}$$

This distortion leads to the same Burgers vector, torsion tensor and dislocation density as that associated with the (k^{I5}) state. A state (k^{I7}) similar to the (k^{I4}) state given by Eq. 6.24 can also be defined as follows:

$$A_K^{k^{I7}} = A_K^{k^{I6}} A_{k^{I6}}^{k^{I7}} \tag{6.45}$$

where

$$A_{k^{I6}}^{k^{I7}} = \begin{bmatrix} \cos\left(\frac{\pi}{2}-\theta\right) & 0 & 0 \\ 0 & \cos\left(\frac{\pi}{2}-\theta\right) & 0 \\ 0 & 0 & 1 \end{bmatrix} \tag{6.46}$$

that when applied to Eq. 6.45 yields

$$\underset{1}{A}_K^{k^{I7}} = \begin{bmatrix} \cos\left(\frac{\pi}{2}-\theta\right) & \sin\left(\frac{\pi}{2}-\theta\right) & 0 \\ -\sin\left(\frac{\pi}{2}-\theta\right) & \cos\left(\frac{\pi}{2}-\theta\right) & 0 \\ 0 & 0 & 1 \end{bmatrix} \tag{6.47}$$

where the expression for $\underset{2}{A}_K^{k^{I7}}$ differs from the above only by an interchange of sign between $\sin\left(\frac{\pi}{2}-\theta\right)$. The grain boundary generated by the distortion given by Eq. 6.45 is shown in Figure 6.6. In spite of the fact that the atomic configuration in Figure 6.6 is identical to that shown in Figure 6.3, the dislocation descriptions of each are entirely different. In particular, the Burgers vector is given by

$$b^{k^{I7}} = -\oint A_K^{k^{I7}} dx^K \tag{6.48}$$

which in view of Eq. 6.45 becomes

$$b^1 = \left\{-4\sin\left(\tfrac{\pi}{2}-\theta\right)\right\}_1 + \left\{-4\sin\left(\tfrac{\pi}{2}-\theta\right)\right\}_2 \tag{6.49a}$$

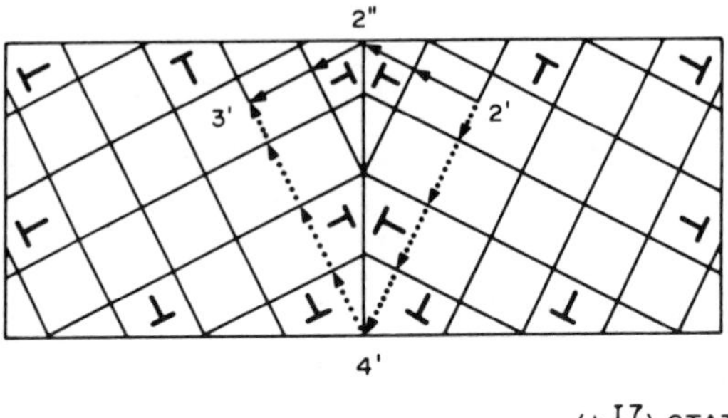

Figure 6.6 Symmetric tilt boundary generated by pure plastic rotation.

which in terms of Figure 6.6 becomes

$$b^1 = \left\{ \underset{4'-3'}{\Delta x^1} \right\}_1 + \left\{ \underset{2'-4'}{\Delta x^1} \right\}_2 \tag{6.49b}$$

The only nonvanishing component of $S_{l^{I7}m^{I7}}^{\cdot\cdot\ k^{I7}}$ is found to be

$$S_{12}^{\cdot\cdot 1} = \left\{ \left(\tfrac{1}{2}\right)\delta(x^1)\sin\left(\tfrac{\pi}{2} - \theta\right) \right\}_1 + \left\{ \left(\tfrac{1}{2}\right)\delta(x^1)\sin\left(\tfrac{\pi}{2} - \theta\right) \right\}_2 \tag{6.50}$$

from which both the Burgers vector and grain boundary dislocation density can be found. We thus have the important finding that the dislocation content of a grain boundary generated by a pure rotation can be described in two alternate ways.

If only the leftmost portion of Figure 5.16*a* undergoes a simple shear, the asymmetric tilt boundary shown in Figure 6.7*a* obtains. It is designated as state (k^{I8}) and can be generated from the (K) state via the following distortion:

$$A_K^{k^{I8}} = \left\{ \underset{1}{A_K^{k^{I8}}}\, H(-x^1) \right\}_1 + \left\{ \delta_K^{k^{I8}} H(+x^1) \right\}_2 \tag{6.51}$$

where

$$\underset{1}{A_K^{k^{I8}}} = \delta_{k^2}^{k^{I8}} \delta_K^L A_L^{k^2} \tag{6.52}$$

and where $A_L^{k^2}$ is given by Eq. 3.16. The closure failures can be obtained from

$$b^{k^{I8}} = -\oint A_K^{k^{I8}} dx^K \tag{6.53a}$$

and

$$\underset{s}{b}^{k^{I8}} = \oint A_K^{k^{I8}} dx^K \tag{6.53b}$$

where the first equation yields

$$b^1 = -\int_3^4 A_2^1 dx^2 = \{4\tan\theta\}_1 = \{2\}_1 \equiv \underset{3'-2''}{\Delta x^1} \tag{6.54}$$

whereas the second gives

$$\underset{s}{b}^2 = \int_3^4 A_2^2 dx^2 + \int_1^2 A_2^2 dx^2 = \{-4\}_1 + \{4\}_2 \tag{6.55}$$

The above equation does not give the closure failure $2''-2'$ shown in Figure 6.7*a*, but must be obtained by first writing

$$\underset{s}{\mathbf{b}} = -4\underset{1}{\mathbf{e}_2} + 4\underset{2}{\mathbf{e}_2} \tag{6.56}$$

where $\underset{1}{\mathbf{e}_2}$ and $\underset{2}{\mathbf{e}_2}$ are the base vectors at the grain boundaries associated with grains #1 and #2, respectively. Note that

$$\left|\underset{1}{\mathbf{e}_2}\right| = \frac{1}{\cos\theta}\left|\underset{2}{\mathbf{e}_2}\right| \tag{6.57}$$

We can also write

$$S_{12}^{\cdot\cdot 1} = \left\{-\tfrac{1}{2}\tan\theta\,\delta(x^1)\right\}_1 \tag{6.58}$$

and

$$\Omega_{12}^{\cdot\cdot 2} = \left\{-\tfrac{1}{2}\delta(x^1)\right\}_1 + \left\{\tfrac{1}{2}\delta(x^1)\right\}_2 \tag{6.59}$$

which can also be used to find b^1 and $\underset{s}{b}^2$ using the surface integration analogue of Eq. 6.53.

The step $2''-2'$ in Figure 6.7*a* can be eliminated by the addition of extra matter within grain #2 to generate the state designated as (k^{I9}) in Figure 6.7*b*. The (k^{I9}) state can also be generated from the (K) state of Figure 5.16*a* via the following distortion:

$$A_K^{k^{I9}} = \left\{\underset{1}{A}_K^{k^{I9}} H(-x^1)\right\}_1 + \left\{\underset{2}{A}_K^{k^{I9}} H(+x^1)\right\}_2 \tag{6.60}$$

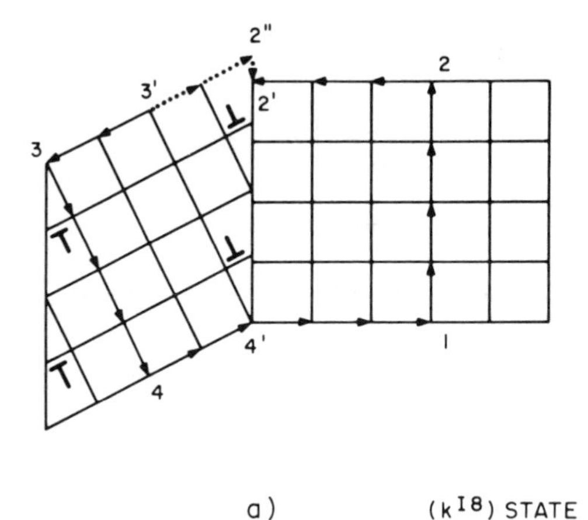

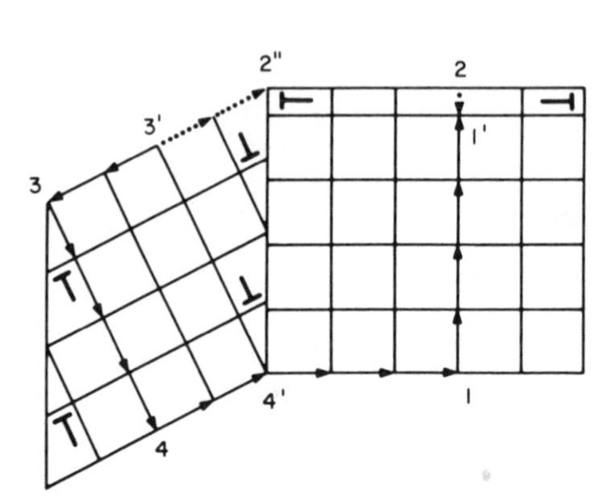

Figure 6.7 Asymmetric tilt boundary. *a*) With free surface ledge *b*) With ledge removed by introduction of dislocation.

where

$$A_K^{k^{I9}} = \delta_{k^{I8}}^{k^{I9}} \delta_K^L A_L^{k^{I8}} \tag{6.61}$$

and where $A_L^{k^{I8}}$ is given by Eq. 6.52, while the nonvanishing components of $A_K^{k^{I9}}$ are
2

$$A_1^1 = A_3^3 = 0 \tag{6.62a}$$

and

$$A_2^2 = \left\{ \frac{1}{\cos\theta} H(+x^1) \right\}_2 \qquad (6.62b)$$

We can next write

$$b^{k^{I9}} = -\oint A_K^{k^{I9}} dx^K \qquad (6.63)$$

from which we obtain

$$b^1 = -\int_3^4 A_2^1 dx^2 = \{4\tan\theta\}_1 = \{2\}_1 \equiv \underset{3'-2''}{\Delta x^1} \qquad (6.64a)$$

and

$$b^2 = -\int_3^4 A_2^2 dx^2 - \int_1^2 A_2^2 dx^2 = \{4\}_1 + \left\{ \frac{4}{\cos\theta} \right\}_2 \equiv \underset{2-1'}{\Delta x^1} \qquad (6.64b)$$

Again, we find $S_{12}^{\cdot\cdot 1}$ to be the same as that given by Eq. 6.58, while

$$S_{12}^{\cdot\cdot 2} = \left\{ -\left(\frac{1}{2}\right)\delta(x^1) \right\}_1 + \left\{ \left(\frac{1}{2}\right)\left(\frac{1}{\cos\theta}\right)\delta(x^1) \right\}_2 \qquad (6.65)$$

Both of these nonvanishing components of the torsion tensor can be used to determine both the grain boundary dislocation Burgers vector and dislocation density. Note that unlike Figure 6.7*a*, the (k^{I9}) state of Figure 6.7*b* possesses no anholonomic object since no new free surfaces are created by the distortion; only dislocations. Finally, in order to round out the present discussion of asymmetric tilt boundaries, Figure 6.7*c* shows an

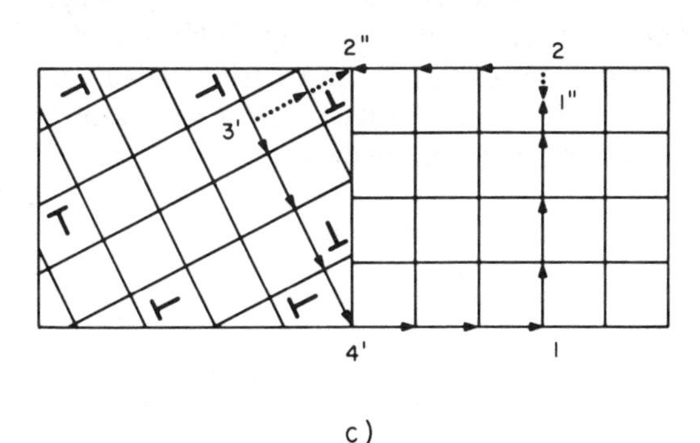

Figure 6.7 (Continued) *c*) Asymmetric tilt boundary generated by plastic rotation of leftmost grain.

asymmetric tilt boundary in which the leftmost grain has undergone a pure plastic rotation. Similar to the case shown in Figure 6.7*b*, dislocations in grain #2 are present. The analysis of this problem follows along the same lines as that given previously.

A second common type of internal surface or interface is that which forms between two different phases. Consider for example the configuration shown in Figure 6.8*a*. It may be viewed as being formed from the K state of Figure 5.16*a* by means of a phase change that simply involves a doubling of the unit cell length in the vertical direction of the material in the leftmost half of the crystal. Unlike the case of Figure 5.16*b*, the length increase is not brought about by an external stress, but by a change in free energy within the crystal, and may thus be viewed as a stress-free strain. However, in order to maintain coherency across the two-phase interface, both crystals must be strained to a common dimension at the interphase boundary and this process leads to the introduction of elastic strains within the body. It is clear from our discussions in Section 5.2, and in particular Figure 5.12, that the horizontal surfaces associated with Figure 6.8*a* are not stress free. However, they can be made so by the introduction of surface dislocations, or more exactly, surface dislocation dipoles, in the manner shown in Figure 6.8*b*. When the fully coherent interface of Figure 6.8 is made fully incoherent, the configuration shown in Figure 6.9*a* obtains. In particular, a set of free surfaces shown by the dotted arrows may be said to have been created in much the same manner as was the case for the completely noncoherent grain boundary of Figure 6.1*b*. It is next possible to eliminate the long-range stresses in Figure 6.8*a* occasioned by the coherency by the introduction of misfit dislocations in the manner shown in Figure 6.9*c*. It is important to note however that the two-phase interface is still fully coherent in much the same way as was the grain boundary of Figure 6.1*a*. Again, however, this boundary may be made fully incoherent to obtain the configuration shown in Figure 6.9*b* which is the two-phase analogue of the incoherent grain boundary of Figure 6.1*b*.

The state shown in Figure 6.9*a* and designated as state (a^I) can be generated from the (K) state of Figure 5.16*a* by means of the distortion $A_K^{a^I}$ that has nonvanishing components

$$A_1^1 = A_3^3 = 1 \tag{6.66a}$$

and

$$A_2^2 = \{H(-x^1)\}_1 + \{H(+x^1)\}_2 \tag{6.66b}$$

Since the base vectors also change, we can write a distortion tensor $B_{a^I}^K$ that is similar to the elastic distortion given in Section 2 but that is stress

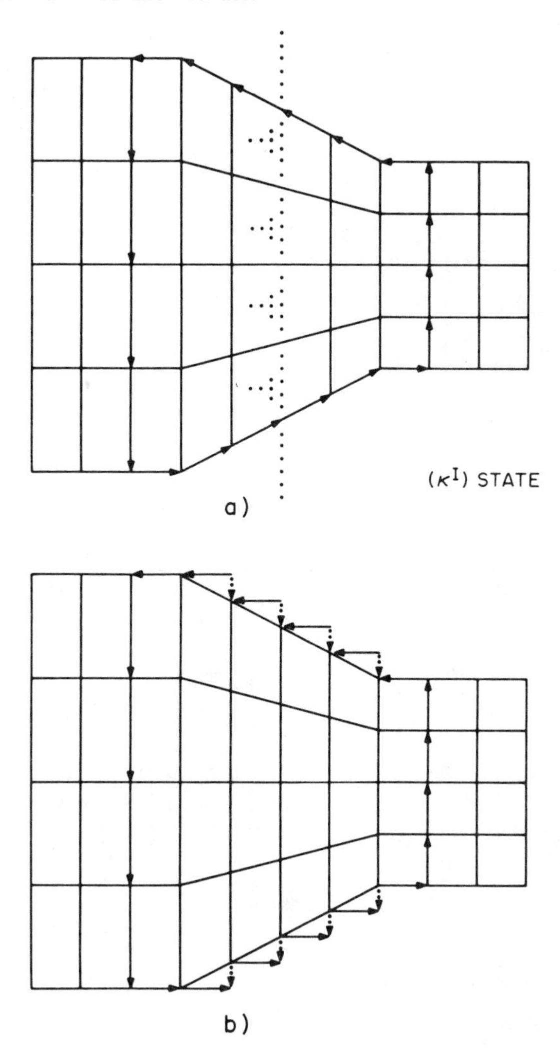

Figure 6.8 Coherent interphase boundary. *a*) Without surface dislocations. *b*) With surface dislocations.

free. In particular, the nonvanishing components are

$$B_1^1 = B_3^3 = 1 \tag{6.67a}$$

and

$$B_2^2 = \{2H(-x^1)\}_1 + \{H(+x^1)\}_2 \tag{6.67b}$$

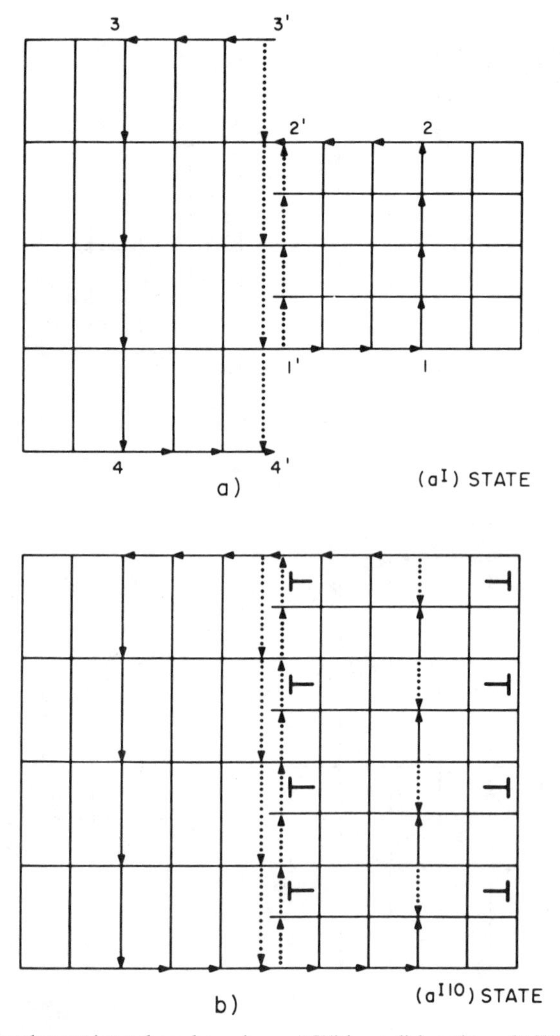

Figure 6.9 Incoherent interphase boundary. *a*) Without dislocations. *b*) With dislocations.

The surface closure failures associated with Figure 6.9*a* may be written as

$$\underset{s}{b}^{a'} = \int A_K^{a'} dx^K \tag{6.68}$$

that in view of Eqs. 6.66 becomes

$$\underset{s}{b}^2 = \{-4\}_1 + \{+4\}_2 \tag{6.69}$$

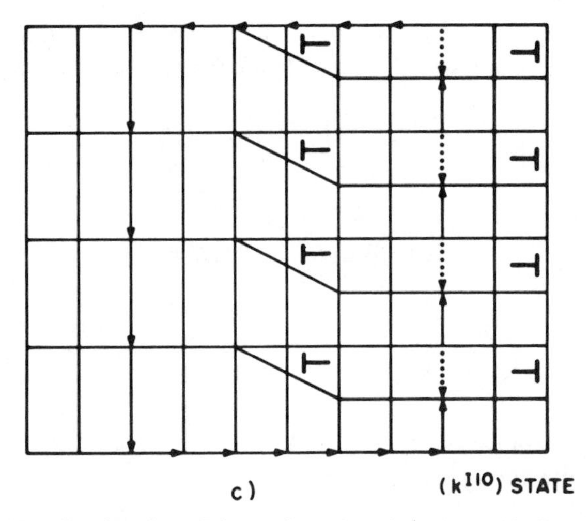

Figure 6.9 (Continued) *c*) Coherent interphase boundary with dislocations.

that in terms of Figure 6.9*a* is simply the dotted vectors

$$\underset{s}{b^2} = \underset{3'-4'}{\Delta x^2} + \underset{1'-2'}{\Delta x^2} \tag{6.70}$$

It is important to note that these vectors cannot be added, but must be drawn separately and associated with their respective phases. In this respect, the two phases are separated along their interface without any loss of generality. The only nonvanishing component of the anholonomic object associated with the (a^I) state is

$$\Omega_{12}^{\cdot\cdot 2} = \left\{ -\tfrac{1}{2}\delta(x^1) \right\}_1 + \left\{ \tfrac{1}{2}\delta(x^1) \right\}_2 \tag{6.71}$$

that can be used to obtain the same result as that given by Eq. 6.69 as well as the newly created free surface density given by

$$\beta^{d^I a^I} = \varepsilon^{d^I b^I c^I} \Omega_{b^I c^I}^{\cdot\cdot\cdot a^I} \tag{6.72}$$

whose only nonvanishing component is

$$\beta^{32} = 2\Omega_{12}^{\cdot\cdot 2} = \{ -\delta(x^1) \}_1 + \{ \delta(x^1) \}_2 \tag{6.73}$$

and simply designates the number of planes of newly created surface divided by the original number of planes prior to the creation of this surface.

Turning now to the dislocated (k^{I10}) state of Figure 6.9*c*, the distortion tensor k^{I10} that generates this state from the (K) state may be written as

$$A_1^1 = A_3^3 = 1 \tag{6.74a}$$

and

$$A_2^2 = \{H(-x^1)\}_1 + \{2H(+x^1)\}_2 \tag{6.74b}$$

while

$$B_{k^{I10}}^K = \delta_{k^{I10}}^{a^I} \delta_L^K B_{a^I}^L \tag{6.75}$$

where $B_{a^I}^L$ is given by Eq. 6.67. The dislocation content associated with the (k^{I10}) state interface may be written as

$$b^{k^{I10}} = -\int A_K^{k^{I10}} dx^K \tag{6.76}$$

that from Eq. 6.74 reduces to

$$b^2 = \{4\}_1 + \{-8\}_2 \equiv \{-4\}_2 \tag{6.77}$$

which is equivalent to the four downward pointing dotted arrows in Figure 6.9*c*. The only component of the torsion tensor associated with the k^{I10} state is given by

$$S_{12}^{\cdot\cdot 2} = \{-(\tfrac{1}{2})\delta(x^1)\}_1 + \{(\tfrac{1}{2})\delta(x^1)\}_2 \tag{6.78}$$

that can also be used to determine the dislocation content of the interphase boundary as well as the interphase boundary dislocation density given by

$$\alpha^{32} = -2S_{12}^{\cdot\cdot 2}/\sqrt{a} \tag{6.79}$$

where a is the determinant of the metric tensor $a_{k^{I10}l^{I10}}$ given by

$$a_{k^{I10}l^{I10}} = A_{k^{I10}}^K A_{l^{I10}}^L a_{KL} \tag{6.80}$$

where again, as in Eq. 4.123, a_{KL} is the metric tensor associated with the (K) state. Equation 6.79 thus yields

$$\alpha^{32} = \{\delta(x^1)\}_1 + \{-2\delta(x^1)\}_2 \equiv -1 \tag{6.81}$$

This number simply corresponds to the four downward pointing dotted arrows in Figure 6.9c divided by the original four planes present within the crystal before the plastic distortion. Similar to the case given by Eq. 4.78, Eq. 6.78 could also have been expressed in terms of a common or dragged area in which case we would need to write

$$S_{12}^{\cdot\cdot 2} = \left\{ -\left(\tfrac{1}{2}\right)\delta(x^1) \right\}_1 + \left\{ 2\left(\tfrac{1}{2}\right)\delta(x^1) \right\}_2 = \left\{ \left(\tfrac{1}{2}\right)\delta(x^1) \right\}_2 \qquad (6.82)$$

It is also important to note that the geometry associated with the (k^{I10}) state of Figure 6.9c is formerly equivalent to that associated with the (κa) state of Figure 5.16b. This means that the misfit dislocation configurations associated with phase changes that result in uniaxial expansion or contraction, simple shear, double shear or a rigid rotation are formerly equivalent to the configurations shown in Figures 5.16b, 5.21a, 5.24a, 5.27a, and 5.29, respectively. It also follows that Figures 5.29a and 6.7c are formerly equivalent.

Figure 6.9b is the anholonomic counterpart of Figure 6.9c. It is clear that the component $S_{12}^{\cdot\cdot 2}$ is the same in both cases. On the other hand, whereas $\Omega_{12}^{\cdot\cdot 2}$ for the (k^{I10}) state was zero, for the (a^{I10}) state of Figure 6.9b we can write

$$\Omega_{12}^{\cdot\cdot 2} = S_{12}^{\cdot\cdot 2} \qquad (6.83)$$

which when substituted into the following equation:

$$b^{a^{I10}} = -\int_S \Omega_{b^{I10}c^{I10}}^{\cdot\cdot a^{I10}} \, dF^{b^{I10}c^{I10}} \qquad (6.84)$$

yields the four upward pointing dotted arrows associated with the four dislocations in Figure 6.9b. There is also an additional contribution to $\Omega_{12}^{\cdot\cdot 2}$ which is the same as that given by Eq. 6.71 and accounts for the nondislocation related closure failures associated with the newly formed boundary surfaces in Figure 6.9b.

The expression given by Eq. 5.41 for the connection is still not its most generalized form. In particular, we could write (Schouten, 1954)

$$\Gamma_{ml}^{k} = \left\{ {k \atop ml} \right\} + S_{ml}^{\cdot\cdot k} - S_{l\cdot m}^{\cdot k} + S_{\cdot ml}^{k} - \Omega_{ml}^{\cdot\cdot k} + \Omega_{l\cdot m}^{\cdot k}$$
$$- \Omega_{\cdot ml}^{k} + \tfrac{1}{2}\left(Q_{ml}^{\cdot\cdot k} + Q_{l\cdot m}^{\cdot k} - Q_{\cdot ml}^{k} \right) \qquad (6.85)$$

When the quantity $Q_{ml}^{\cdot\cdot k}$ vanishes, the connection is called metric, which simply means that a vector does not change its length upon parallel displacement. This follows from the fact that (Schouten, 1954)

$$\nabla_m g_{lk} = -Q_{mlk} \qquad (6.86)$$

The $Q_{ml}^{\cdot\cdot k}$ tensor can be related to the stress-free strain tensor e_{ln} as follows (deWit, 1978; Anthony, 1970, 1971; Marcinkowski, 1977c):

$$Q_{ml}^{\cdot\cdot k} = -2g^{kn}\partial_m e_{ln} \tag{6.87}$$

Now it is obvious that the (k) state of Figure 4.2 and the (κ^I) state of Figure 6.8*a* are closely related to one another. In the former case we can write

$$\Gamma_{[ml]}^{k} = S_{ml}^{\cdot\cdot k} \tag{6.87a}$$

whereas in the latter

$$\Gamma_{[\mu^I\lambda^I]}^{\kappa^I} = \tfrac{1}{2} Q_{\mu^I\lambda^I}^{\cdot\cdot\kappa^I} \tag{6.87b}$$

In other words, a virtual dislocation content (Marcinkowski, 1970) can be associated with the (κ^I) state that is related to a varying metric tensor. For the stress-free strain, we can write

$$e_{\lambda^I\eta^I} = \tfrac{1}{2}(g_{\lambda^I\eta^I} - \delta_{\lambda^I\eta^I}) \tag{6.88}$$

Since the nonvanishing components of $g_{\lambda^I\eta^I}$ are

$$g_{11} = g_{33} = 1 \tag{6.89a}$$

and

$$g_{22} = \{4H(-x^1)\}_1 + \{H(+x^1)\}_2 \tag{6.89b}$$

we have

$$e_{22} = \tfrac{1}{2}[\{4H(-x^1)\}_1 + \{H(+x^1)\}_2 - 1] \tag{6.90}$$

The only nonvanishing component of $Q_{\mu^I\lambda^I}^{\cdot\cdot\kappa^I}$, according to Eq. 6.87, can be written as

$$Q_{12}^{\cdot\cdot 2} = -2g^{22}\partial_1 e_{22} = -g^{22}[\{-4\delta(x^1)\}_1 + \{\delta(x^1)\}_2] \tag{6.91}$$

Also since

$$g^{\kappa^I\eta^I} g_{\eta^I\lambda^I} = \delta_{\lambda^I}^{\kappa^I} \tag{6.92}$$

we can write

$$g^{11} = g^{33} = 1 \tag{6.93a}$$

and

$$g^{22} = \left\{\tfrac{1}{4}H(-x^1)\right\}_1 + \left\{H(+x^1)\right\}_2 \tag{6.93b}$$

so that Eq. 6.91 becomes

$$Q_{12}^{\cdot\cdot 2} = \{\delta(x^1)\}_1 + \{-\delta(x^1)\}_2 \tag{6.94}$$

This quantity can in turn be substituted into the following equation:

$$b^{\kappa'} = \int_s \Gamma_{[\mu'\lambda']}^{\kappa'} dF^{\mu'\lambda'} = \frac{1}{2}\int_s Q_{\mu'\lambda'}^{\cdot\cdot\kappa'} dF^{\mu'\lambda'} \tag{6.95}$$

to obtain

$$b^2 = \{4\}_1 + \{-4\}_2 \tag{6.96}$$

or in vector form

$$\mathbf{b} = b^{\kappa'}\mathbf{e}_{\kappa'} = \{8\mathbf{e}_2\}_1 + \{-4\mathbf{e}_2\}_2 = \{4\mathbf{e}_2\}_1 \tag{6.97}$$

This is the same result as that given by Eq. 4.75b which was obtained for the (k') state of Figure 4.2. Thus, in spite of the fact that the (κ^I) state is purely elastic, we may nevertheless associate a virtual dislocation content with it of the same strength as that given for the (k) state of Figure 4.2. These virtual dislocations are drawn dotted in Figure 6.8*a*. It is also possible to define a dislocation density tensor for the (κ^I) state that can be written as

$$\alpha^{\eta'\kappa'} = \tfrac{1}{2}\varepsilon^{\eta'\mu'\lambda'} Q_{\mu'\lambda'}^{\cdot\cdot\kappa'} \tag{6.98}$$

that has a nonvanishing component

$$\alpha^{32} = Q_{12}^{\cdot\cdot 2}/\sqrt{a} = \{2\}_1 + \{-1\}_2 \tag{6.99}$$

A coherent interphase boundary can also be generated by a phase transformation that results in a simple shear such as shown in Figure 6.10 and denoted as state (κ^{I2}). Figure 6.10*b* shows this same state, but in terms

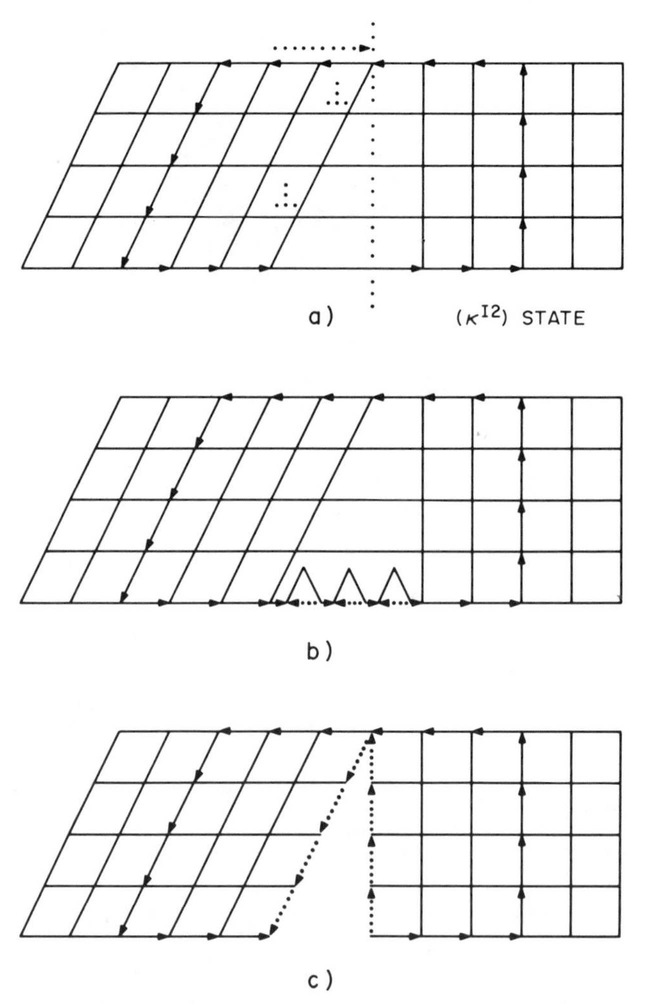

Figure 6.10 Coherent interphase boundary formed by simple shear. *a*) Without surface dislocations. *b*) With surface dislocations. *c*) Noncoherent counterpart of *a*).

of surface dislocations, whereas Figure 6.10*c* again shows this same state, but after the surface dislocations have annihilated with one another to generate a fully noncoherent boundary. Again, there is no actual dislocation associated with the (κ^{I2}) state of Figure 6.10*a*; however, a virtual dislocation content may be associated with it by writing

$$Q_{\dot{\mu}^{I2}\dot{\lambda}^{I2}}{}^{\kappa^{I2}} = -2g^{\kappa^{I2}\eta^{I2}}\partial_{\mu^{I2}}e_{\lambda^{I2}\eta^{I2}} \tag{6.100}$$

where the strain tensor is given by

$$e_{\lambda^{12}\eta^{12}} = \tfrac{1}{2}\left(g_{\lambda^{12}\eta^{12}} - \delta_{\lambda^{12}\eta^{12}}\right) \tag{6.101}$$

and where

$$g_{\lambda^{12}\eta^{12}} = \left\{ \underset{1}{g}_{\lambda^{12}\eta^{12}} H(-x^1) \right\}_1 + \left\{ \delta_{\lambda^{12}\eta^{12}} H(+x^1) \right\}_2 \tag{6.102}$$

where

$$\underset{1}{g}_{\lambda^{12}\eta^{12}} = \delta^{\kappa^2}_{\lambda^{12}} \delta^{\lambda^2}_{\mu^{12}} g_{\kappa^2\lambda^2} \tag{6.103}$$

where $g_{\kappa^2\lambda^2}$ is given by Eq. 2.25. The only nonvanishing component of Eq. 6.100 thus becomes

$$Q_{12}^{\cdot\cdot 1} = -2g^{11}\partial_1 e_{21} - 2g^{12}\partial_1 e_{22} \tag{6.104}$$

We can also write

$$g^{\lambda^{12}\eta^{12}} = \left\{ \underset{1}{g}^{\lambda^{12}\eta^{12}} H(-x^1) \right\}_1 + \left\{ \delta^{\lambda^{12}\eta^{12}} H(+x^1) \right\}_2 \tag{6.105}$$

where

$$\underset{1}{g}^{\lambda^{12}\eta^{12}} = \begin{bmatrix} \tan^2\theta + 1 & -\tan\theta & 0 \\ -\tan\theta & 1 & 0 \\ 0 & 0 & 1 \end{bmatrix} \tag{6.106}$$

Equation 6.104 thus becomes

$$Q_{12}^{\cdot\cdot 1} = \left\{ \tan\theta\, \delta(x^1) \right\}_1 \tag{6.107}$$

that when substituted into

$$b^{\kappa^{12}} = \frac{1}{2} \int_S Q_{\mu^{12}\lambda^{12}}^{\cdot\cdot\ \kappa^{12}}\, dF^{\mu^{12}\lambda^{12}} \tag{6.108}$$

yields

$$b^1 = \{4\tan\theta\}_1 = \{2\}_1 \tag{6.109}$$

that corresponds to the virtual dislocation content of Figure 6.10*a*. It in effect corresponds to the distance shown by the dotted arrow. Equation 6.107 can also be used to determine the quasi-dislocation density for the (κ^{I2}) state of Figure 6.10*a* to be

$$\alpha^{31} = Q_{12}^{\cdot\cdot 1} = \{\tan\theta\}_1 \tag{6.110}$$

Since the (a^I) state shown in Figure 6.9*a* is metric, $Q_{a^I b^I}^{\cdot\cdot c^I} = 0$. However, it could be made nonmetric by writing

$$Q_{a^I b^I}^{\cdot\cdot c^I} = \delta_{a^I}^{\mu^I} \delta_{b^I}^{\lambda^I} \delta_{\kappa^I}^{c^I} Q_{\mu^I \lambda^I}^{\cdot\cdot \kappa^I} \tag{6.111}$$

where the only nonvanishing component of $Q_{\mu^I \lambda^I}^{\cdot\cdot \kappa^I}$ is given by Eq. 6.94. It also follows from Eq. 6.71 that

$$Q_{a^I b^I}^{\cdot\cdot c^I} = -2\Omega_{a^I b^I}^{\cdot\cdot c^I} \tag{6.112}$$

Thus we could write

$$b^{c^I} = \int_s \left[\frac{1}{2} Q_{a^I b^I}^{\cdot\cdot c^I} + \Omega_{a^I b^I}^{\cdot\cdot c^I} \right] dF^{a^I b^I} = 0 \tag{6.113}$$

This simply means that the quasi-dislocation content, measured by $Q_{a^I b^I}^{\cdot\cdot c^I}$, is just compensated by the newly created free surfaces measured by $\Omega_{a^I b^I}^{\cdot\cdot c^I}$. Thus we have a situation similar to that first encountered in Eq. 5.44, except that now the $Q_{a^I b^I}^{\cdot\cdot c^I}$ tensor replaces the torsion tensor. Similar arguments can also be used for the simple shear configuration of Figure 6.10*c*. The counterpart of the (a^{I10}) state of Figure 6.9*b* can be obtained for Figure 6.10*c* by filling the void between the dotted arrows with matter.

If the leftmost portion of the (K) state of Figure 5.16*a* undergoes a distortion similar to that obtained for the (κ^4) state of Figure 2.3*a*, the resulting configuration is similar to that shown in Figure 6.8*a*. The reason for this is that from Eqs. 2.31 and 2.32 we can write

$$g_{22} = \left\{ \left(\frac{1}{\cos^2\theta} \right) H(-x^1) \right\}_1 + \{H(+x^1)\}_2 \tag{6.114}$$

and

$$e_{22} = \frac{1}{2} \left[\left\{ \left(\frac{1}{\cos^2\theta} \right) H(-x^1) \right\}_1 + \{H(+x^1)\}_2 - 1 \right] \tag{6.115}$$

which when substituted into an equation of the form given by Eq. 6.91 leads to the same result as that given by Eq. 6.94. In the case where the left half of the (K) state in Figure 5.16a undergoes a rigid rotation, the right side simply follows along with it so that no resultant elastic distortion obtains which means that $Q_{ml}^{\cdot\cdot k}$ for this case vanishes. Finally, for completeness, it is apparent that if the leftmost portion of Figure 5.16a undergoes a phase change which involves a distortion of the type given by the (κ^6) state of Figure 2.4, then the interface between the two phases consists of two distinct sets of virtual dislocations; one set similar to those shown in Figure 6.8a, and the other set similar to those shown in Figure 6.10a. This is apparent from the nature of the metric and strain tensors given by Eqs. 2.39 and 2.40, respectively. All of the virtual dislocation arrays considered thus far could also have been generated by subjecting the leftmost region of Figure 5.16a to some suitable stress which in turn generated an elastic distortion within this region. In fact, the distortion in this region could be brought about in innumerable ways, that is, temperature gradient, electrical field gradient, magnetic field gradient, and so forth, that in turn generates thermal, piezoelectric, magnetostrictive, and so forth, stresses. Our analysis of free surfaces and interphase boundaries is thus essentially complete in that we have considered boundaries separating elastic-plastic, plastic-plastic, elastic-undistorted and plastic-undistorted. The analysis of boundaries separating regions of differing elastic distortion is similar to that associated with boundaries separating elastic from undistorted regions.

For completeness let us consider a grain boundary between two phases of differing lattice constant such as is shown in Figure 6.11 and denoted as state (k^{I11}). Such a state can be generated from the (K) reference state by means of the distortion $A_K^{k^{I11}}$ whose nonvanishing components are given by

$$A_1^1 = A_2^2 = A_3^3 = \left\{\left(\tfrac{5}{4}\right)H(-x^1)\right\}_1 + \left\{H(+x^1)\right\}_2 \tag{6.116a}$$

and

$$A_2^1 = \left\{\left(\tfrac{5}{4}\right)H(-x^1)\tan\theta\right\}_1 + \left\{-H(+x^1)\tan\theta\right\}_2 \tag{6.116b}$$

The grain boundary dislocation content can be obtained from the following line integral:

$$b^{k^{I11}} = -\oint A_K^{k^{I11}}\, dx^K \tag{6.117}$$

which has nonvanishing components

$$b^1 = \left\{\left(\tfrac{5}{4}\right)4\tan\theta\right\}_1 + \left\{4\tan\theta\right\}_2 = \left\{\tfrac{5}{2}\right\}_1 + \{2\}_2 \tag{6.118a}$$

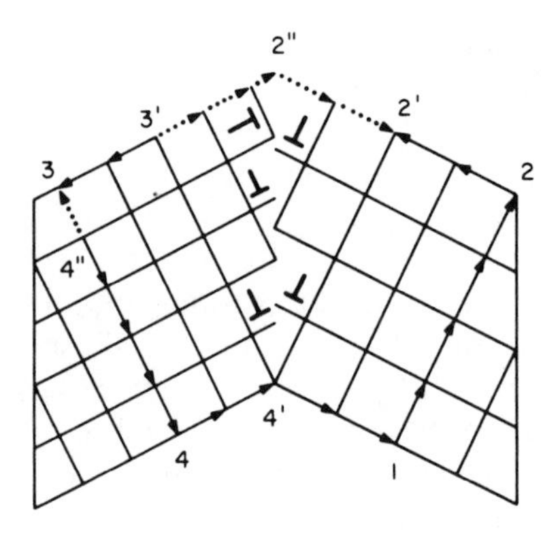

(k^{III}) STATE

Figure 6.11 Grain boundary separating two phases of differing lattice constant.

or in terms of Figure 6.11

$$b^1 = \underset{3'-2''}{\Delta x^1} + \underset{2''-2'}{\Delta x^1} \tag{6.118b}$$

while the other component yields

$$b^2 = \left\{\left(\tfrac{5}{4}\right)4\right\}_1 + \{-4\}_2 = \{+1\}_1 \equiv \underset{4''-3}{\Delta x^2} \tag{6.119}$$

The surface integral counterpart of Eq. 6.118 is

$$b^{k^{III}} = -\int_S S_{l^{III}m^{III}}^{\cdot\cdot\,k^{III}}\, dF^{l^{III}m^{III}} \tag{6.120}$$

where the two nonvanishing components of the torsion tensor are found to be

$$S_{12}^{\cdot\cdot 1} = \left\{-\left(\tfrac{1}{2}\right)\left(\tfrac{5}{4}\right)\tan\theta\left(\tfrac{4}{5}\right)^2\delta(x^1)\right\}_1 + \left\{-\left(\tfrac{1}{2}\right)\tan\theta\,\delta(x^1)\right\}_2 \tag{6.121a}$$

and

$$S_{12}^{\cdot\cdot 2} = \left\{-\left(\tfrac{1}{2}\right)\left(\tfrac{5}{4}\right)\left(\tfrac{4}{5}\right)^2\delta(x^1)\right\}_1 + \left\{\left(\tfrac{1}{2}\right)\delta(x^1)\right\}_2 \tag{6.121b}$$

The $\delta(x^1)$ in the above relations are in terms of the (K) state coordinates, but can be expressed in terms of (k^{III}) state coordinates by writing, analogous to Eq. 4.108,

$$\delta(x^{K}) = \delta(x^{k^{I14}}) A_{K}^{k^{I14}} \tag{6.122a}$$

or

$$\delta(x^{1}) = \delta(x^{1})\left(\tfrac{5}{4}\right) \tag{6.122b}$$

When substituted into Eq. 6.120, Eq. 6.121 lead to the same results as those given by Eqs. 6.118 and 6.119. Finally, for the grain boundary dislocation density, we may write

$$\alpha^{n^{I11} k^{I11}} = -\varepsilon^{n^{I11} l^{I11} m^{I11}} S_{l^{I11} m^{I11}}^{\cdot \quad \cdot \quad k^{I11}} \tag{6.123}$$

which gives the two components

$$\alpha^{31} = -2S_{12}^{\cdot\cdot 1}/\sqrt{a} = \left\{\delta(x^{1})\left(\tfrac{5}{4}\right)\tan\theta\right\}_{1} + \left\{\delta(x^{1})\tan\theta\right\}_{2} \tag{6.124a}$$

and

$$\alpha^{31} = -2S_{12}^{\cdot\cdot 2}/\sqrt{a} = \left\{\delta(x^{1})\left(\tfrac{5}{4}\right)\right\}_{1} + \left\{\delta(x^{1})\right\}_{2} \equiv \left\{\tfrac{1}{4}\right\}_{1} \tag{6.124b}$$

The quantity a in the above equations has the value (4/5) in grain #1 and unity in grain #2.

REVIEW

It was shown in Chapter 4 how a given volume of a crystal could be plastically deformed and that the deformed and undeformed regions would be separated from each other by some suitable planar array of dislocations. More generally, adjacent regions of a crystal may be subjected to different deformations. In the specific case of Figure 6.1a, when two regions of a crystal are sheared by the same magnitude but in opposite directions, a symmetric type grain boundary is obtained. Once again the distortion tensor A_K^k may be written in terms of Heaviside-type step functions that describe the separate distortions in the individual grains. The particular distortion associated with Figure 6.1a is given by Eq. 6.1. It also follows that the distortion tensor A_K^k can be integrated about a given path, as per Eq. 6.2, to obtain the grain boundary dislocation content. Furthermore, the distortion tensor can be used to obtain a grain boundary torsion tensor that can in turn be integrated with respect to the area enclosed within the

Burgers circuit so as to obtain an alternate way of determining the grain boundary dislocation content. This particular method is indicated in Eq. 6.4. Finally, the grain boundary dislocation content can be obtained from the torsion tensor in accordance with Eq. 6.7.

It is interesting to note that the grain boundary of Figure 6.1*a* possesses very high elastic strains. These strains, however, can be completely removed by the perfect tearing process that leads to the grain boundary configuration shown in Figure 6.1*b*. Associated with the tearing process is the creation of new surface. In particular, steps associated with the grain boundary dislocation are formed. These steps are equivalent to that which is obtained when the internal dislocations reach the surface of the crystal, as described in Chapter 5. More specifically, the state shown in Figure 6.1*b* may be termed an anholonomic state possessing an anholonomic object $\Omega_{cb}^{\cdot\cdot a}$ that just equals that of the torsion tensor as expressed by Eq. 6.15. Thus, when the area integration of these two quantities is carried out in accordance with Eq. 6.14, the Burgers vectors associated with the grain boundary dislocations is just compensated by the extra length of newly created free surface steps or ledges.

In effect, Figures 6.1*a* and 6.1*b* represent the two extreme cases of grain boundary morphology. In the former configuration the elastic strain energy is a maximum whereas that energy associated with grain boundary free surfaces is zero. On the other hand, the strain energy associated with the configuration in Figure 6.1*b* is zero whereas the grain boundary free surface energy is a maximum. In reality, an actual grain boundary has a configuration that lies somewhere in between, such as depicted by Figure 7.5 in the following chapter. This very important case is considered in greater detail in Chapter 13 where the most precise grain boundary energy calculation given to date is presented. The drawback associated with most of the grain boundary models now in existence is that they are rigid models of the type given by the torn state shown in Figure 6.1*b* and can thus say nothing of the strain energy present in real boundaries.

A second important type of internal boundary is that associated with a two-phase interface such as is shown in Figure 6.8*a*. Such a boundary may be viewed as being formed by a phase change that causes the volume within the leftmost portion of the body in Figure 6.8*a* to increase. The displacements across the boundary are continuous which means that the interface is fully coherent. It is important to note that the volume change arises from a stress-free strain $e_{\kappa\lambda}$ and could also have arisen in a number of other different ways such as through a temperature gradient. It is apparent that no real dislocations, as evidenced by extra half planes, exist for the state shown in Figure 6.8*a*. However, a quasi-dislocation content can be associated with this boundary as indicated by the dotted dislocation symbols. These quasi dislocations have associated with them a tensor $Q_{\mu\lambda}^{\cdot\cdot\kappa}$ of the form given by Eq. 6.87. In particular, it is related to the way that the stress-free strain, or equivalently the metric tensor, varies with distance across the interphase boundary. The quasi-dislocation content associated with the boundary can be obtained by intergrating $Q_{\mu\lambda}^{\cdot\cdot\kappa}$ with respect to area in accordance with Eq. 6.95. A state characterized by a nonvanishing $Q_{\mu\lambda}^{\cdot\cdot\kappa}$ is termed a nonmetric state. This is an important result in that it allows stress-free internal distortions to be represented in terms of quasi or virtual dislocations. We see, however, in the next chapter that a still more powerful way of viewing these stress-free distortions is in terms of a common or coincidence lattice that allows the dislocation concept to be used in its entirety.

CHAPTER 7

Coincidence Site Lattice

We have seen in Chapter 2 that an elastic distortion is characterized by a dragging of the coordinates and a change in the metric tensor. Chapter 3, on the other hand, showed that a plastic distortion was characterized by a dragging of the metric tensor with a corresponding change in coordinates. Furthermore, Chapter 4 showed the close relationship between plastic deformation and dislocations. Finally, in Chapter 5, it was shown that any state of elastic distortion could be described in terms of some suitable distribution of dislocations. The generalized treatment by which this transformation may be carried out involves the concept of a coincidence site lattice and forms the subject of the present chapter.

Reference to Figure 6.1*d* shows that a number of points associated with each of the two grains are common to one another at the grain boundary. These points can be used to define a coincidence site lattice common to both grains (Marcinkowski and Sadananda, 1975). Figure 7.1*a* shows the reconstruction of Figure 6.1*d* in terms of such a lattice, where the unit cell dimension is given as a_{oc}. For clarity, the original lattice is shown with respect to only two of the coincidence site lattice unit cells. The coincidence site lattice is also constructed so that one of its unit cell edges is coincident with the boundary. Figure 7.1*b* shows that it is also possible to subdivide the coincidence site lattice into a still smaller lattice of unit cell size a_{ocs} and is termed the coincident site lattice sublattice. This sublattice is outlined by dotted lines in Figure 7.1*b*. Now it is clear from the geometry of Figure 7.1*a* that

$$a_{oc} = \sqrt{N^2 + M^2}\ a_o = Ma_o/\cos\theta \tag{7.1}$$

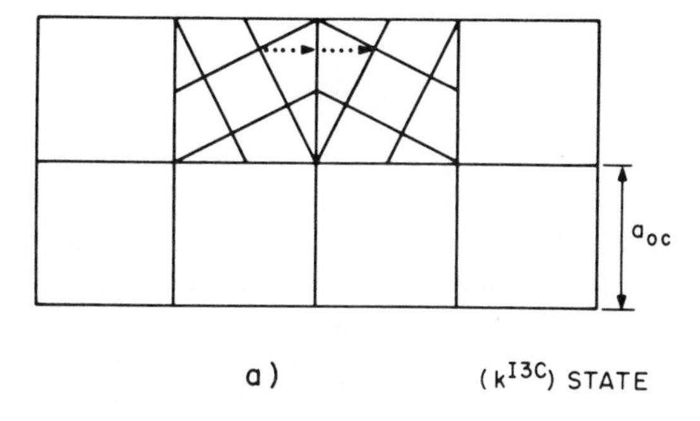

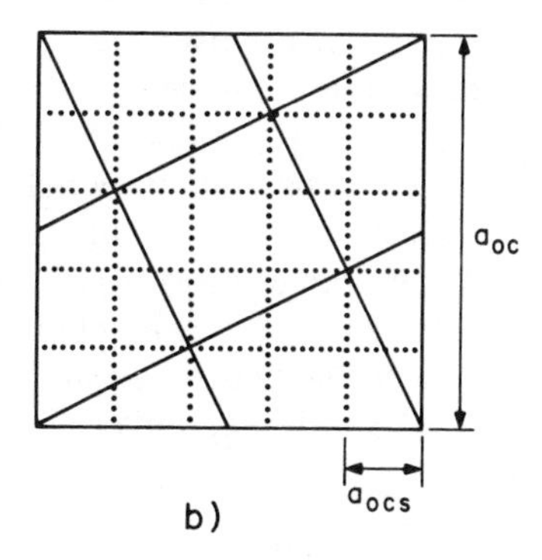

Figure 7.1 *a*) Coincidence site lattice associated with the symmetric grain boundary of Figure 6.1*d*. *b*) Coincidence site sublattice associated with *a*).

where a_o is the crystal lattice unit cell size while N and M are integers defined by

$$\tan\theta = \frac{N}{M} \tag{7.2}$$

It is also clear from Figure 7.1*a* that $N=1$, while $M=2$. It also follows from Eq. 7.1 and from the fact that

$$a_o = \sqrt{N^2+M^2}\; a_{ocs} \tag{7.3}$$

while

$$a_{ocs} = \frac{a_{oc}}{N^2+M^2} \tag{7.4a}$$

so that in the particular case shown in Figure 7.1*b*

$$a_{ocs} = \frac{a_{oc}}{5} \tag{7.4b}$$

Figure 7.2 shows the coincidence site lattice associated with the interphase boundary of Figure 6.9*c*. In this case, the coincidence site lattice is outlined by the solid lines and has unit cell dimensions given by $a_{oc} = a_o$ and $b_{oc} = 2a_o$. The dotted and solid line combination, on the other hand, delineates the coincidence site lattice sublattice which in turn corresponds to the unit cell dimensions of the crystal lattice of the rightmost phase. Dislocations that can be described in terms of the coincidence site lattice are still present in Figure 7.2. However, although they generate local distortions in the vicinity of the interface, they give rise to no long-range distortions since they occur in pairs of opposite sign. It is this absence of long-range distortions that allows the coincidence site lattice to be constructed. It is also clear from inspection of Figure 6.11 that a coincidence site lattice can be constructed for a grain boundary between two different phases or for that matter between any type of interphase boundary, in general, that does not possess long-range distortions. Another interesting rule is that any asymmetric boundary of misorientation angle ϕ can be described in terms of the coincidence site lattice as shown dotted in Figure 7.3. In particular, the coincidence site lattice unit cell associated with the

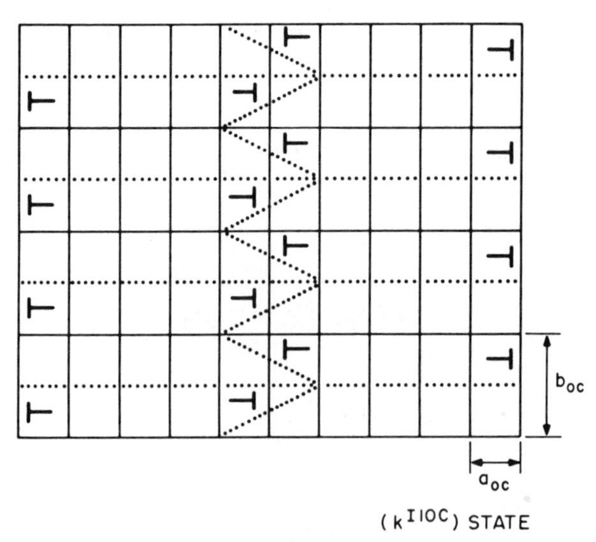

Figure 7.2 Coincidence site lattice associated with the interphase boundary of Figure 6.9*b*.

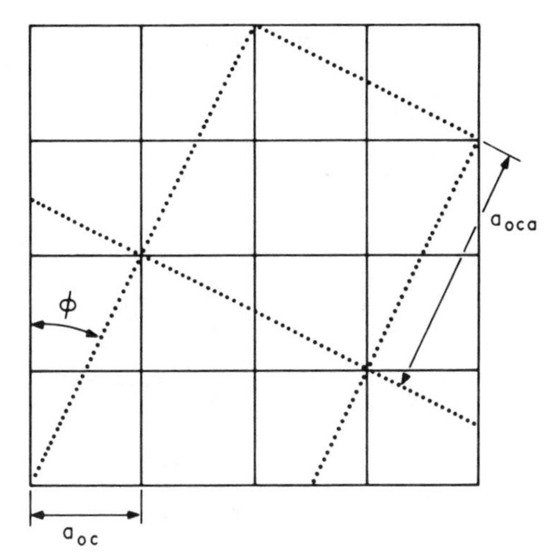

Figure 7.3 Coincidence site lattice construction associated with an asymmetric boundary of misorientation angle ϕ.

asymmetric boundary is given by

$$a_{oca} = \sqrt{o^2 + p^2}\; a_{oc} \tag{7.5}$$

where

$$\tan\theta = \frac{o}{p} \tag{7.6}$$

In the particular case shown in Figure 7.3, $o = 1$ and $p = 2$. The similarity of the last two equations with those of Eqs. 7.1 and 7.2 is obvious. In addition, a coincidence site lattice sublattice can also be associated with Figure 7.3 and is given by

$$a_{ocas} = \frac{a_{oca}}{o^2 + p^2} \tag{7.7}$$

The coincidence site lattice designated as state (k^{I10c}) in Figure 7.2 can be obtained from the (k^{I10}) state shown in Figure 6.9c by means of the coordinate transformation $C_{k^{I10}}^{k^{I10c}}$ which yields

$$A_K^{k^{I10c}} = C_{k^{I10}}^{k^{I10c}} A_K^{\kappa^I} A_{\kappa^I}^{k^{I10}} = C_{k^{I10}}^{k^{I10c}} \delta_K^{\kappa^I} A_{\kappa^I}^{k^{I10}} = \delta_K^{k^{I10c}} \tag{7.8}$$

where $A_K^{k^{I10}}$ is given by Eq. 6.74 and where

$$C_{k^{I10}}^{k^{I10c}} = \left\{ \delta_{k^{I10}}^{k^{I10c}} H(-x^1) \right\}_1 + \left\{ \delta_{k^{I10}}^{K} \delta_{\lambda}^{k^{I10c}} B_K^{\lambda} H(+x^1) \right\}_2 \tag{7.9}$$

that has nonvanishing components

$$C_1^1 = C_3^3 = 1 \tag{7.10a}$$

and

$$C_2^2 = \left\{ H(-x^1) \right\}_1 + \left\{ \left(\tfrac{1}{2}\right) H(+x^1) \right\}_2 \tag{7.10b}$$

In addition

$$B_K^{k^{I10c}} = C_{k^{I10}}^{k^{I10c}} B_K^{k^{I10}} = C_{k^{I10}}^{k^{I10c}} \delta_K^{k^{I10}} \tag{7.11}$$

Comparison of these results with those of Eqs. 3.38–3.40 shows that the construction of a coincidence site lattice is analogous to expressing a plastic distortion in terms of an elastic distortion. It is therefore obvious that

$$b^{k^{I10c}} = -\oint A_K^{k^{I10c}} dx^K = 0 \tag{7.12}$$

In particular, the dislocation content of a boundary vanishes when the boundary is expressed in terms of the coincidence site lattice.

In a similar way, the coincidence site lattice labeled as state (k^{I3c}) in Figure 7.1*a* can be obtained from the (k^{I3}) state of Figure 6.1*d* by means of the coordinate transformation

$$C_{k^{I3}}^{k^{I3c}} = \left\{ \left(\frac{1}{M} \right) \delta_{k^{I3}}^{K} \delta_{\lambda^4}^{k^{I3c}} \underset{1}{B_K^{\lambda^4}} H(-x^1) \right\}_1 + \left\{ \left(\frac{1}{M} \right) \delta_{k^{I3}}^{K} \delta_{\lambda^4}^{k^{I3c}} \underset{2}{B_K^{\lambda^4}} H(+x^1) \right\}_2 \tag{7.13}$$

which yields

$$A_K^{k^{I3c}} = C_{k^{I3}}^{k^{I3c}} A_K^{k^{I3}} = \left(\frac{1}{M} \right) \delta_K^{k^{I3c}} \tag{7.14}$$

while

$$B_K^{k^{I3c}} = C_{k^{I3}}^{k^{I3c}} B_K^{k^{I3}} = C_{k^{I3}}^{k^{I3c}} \delta_K^{k^{I3}} \tag{7.15}$$

Again, except for a factor of $(1/M)$, the results are similar to those given by Eqs. 7.8–7.11. If the coincidence site lattice sublattice such as shown in Figure 7.1*b* is to be chosen, then $(1/M)$ in Eq. 7.13 must be replaced by $\sqrt{N^2+M^2}\,/M$. We can also write

$$b^{k^{I3c}} = -\oint A_K^{k^{I3c}}\,dx^K = 0 \tag{7.16}$$

Once again, the perfect coincidence site lattice provides a reference frame that enables the dislocation content of the body to vanish. It should be pointed out again here that the transformations given by Eqs. 7.9 and 7.13 are coordinate transformations and thus have no strains associated with them. In addition, the base vectors are also changed by such a transformation. In particular

$$\mathbf{e}_{k^{I3c}} = B_{k^{I3c}}^{K}\mathbf{e}_K \tag{7.17}$$

where $\mathbf{e}_{k^{I3c}}$ is the base vector associated with the coincidence site lattice whose magnitude is given by Eq. 7.1.

It is now possible to represent the tensor quantities associated with the (k^{I3}) state in terms of the coordinate system associated with the coincidence site lattice, that is, (k^{I3c}). In particular, we can write for the Burgers vector

$$b^{k^{I3c}} = C_{k^{I3}}^{k^{I3c}} b^{k^{I3}} \tag{7.18}$$

where $C_{k^{I3}}^{k^{I3c}}$ is given by Eq. 7.13. Equation 7.18 thus yields

$$b^1 = C_1^1 b^1 = \left\{\frac{\cos^2\theta}{M}b^1\right\}_1 + \left\{\frac{\cos^2\theta}{M}b^1\right\}_2 \tag{7.19a}$$

or

$$b^1 = \left\{\tfrac{2}{5}\right\}_1 + \left\{\tfrac{2}{5}\right\}_2 \tag{7.19b}$$

This result corresponds to the resolution of the Burgers vector of the crystal lattice dislocations into the coordinate system of the coincidence site lattice. This is illustrated by the pair of dotted arrows in Figure 7.1*a*. In terms of the coincidence site lattice sublattice of Figure 7.1*b*, Eq. 7.19a becomes

$$b^1 = \left\{\frac{\sqrt{N^2+M^2}}{M}\cos^2\theta b^1\right\}_1 + \left\{\frac{\sqrt{N^2+M^2}}{M}\cos^2\theta b^1\right\}_2 \tag{7.20a}$$

or

$$b^1 = \{2\}_1 + \{2\}_2 \tag{7.20b}$$

It is to be noted here that Eq. 7.18 does not contradict that given by Eq. 7.12. In the former case, the components of a tensor quantity are simply changed to that of another coordinate system whereas in the latter case, the nature of the way the distortion itself is represented is changed. The quantities given by Eqs. 7.19 and 7.20 are simply the vector sums of those shown in Figure 6.1*d*, that is,

$$\mathbf{b}_{GB} = \mathbf{b}_1 + \mathbf{b}_2 \tag{7.21}$$

where the subscript *GB* denotes grain boundary dislocation (Marcinkowski, Sadananda, Tseng, 1973), while the subscripts 1 and 2 denote the crystal lattice dislocations associated with each of the individual grains.

The torsion tensor associated with the (k^{I3c}) state can also be found from the following relation:

$$S_{l^{I3c}q^{I3c}}^{\cdot\ \cdot\ k^{I3c}} = C_{l^{I3c}}^{l^{I3}} C_{m^{I3c}}^{m^{I3}} C_{k^{I3}}^{k^{I3c}} S_{l^{I3}m^{I3}}^{\cdot\ \cdot\ k^{I3}} \tag{7.22}$$

which has a nonvanishing component

$$S_{12}^{\cdot\cdot 1} = \bar{C}_1^1 \bar{C}_2^2 C_1^1 S_{12}^{\cdot\cdot 1} - \bar{C}_1^2 \bar{C}_2^1 C_1^1 S_{12}^{\cdot\cdot 1} \tag{7.23a}$$

where in view of Eqs. 7.13 and 6.5 we obtain

$$S_{12}^{\cdot\cdot 1} = \left\{ -M^2 \frac{\cos\theta}{M} \left(\frac{1}{2}\right) \left(\frac{\cos\theta}{M} \delta(x^1)\right) \tan\theta \right\}_1$$

$$+ \left\{ -M^2 \frac{\cos\theta}{M} \left(\frac{1}{2}\right) \left(\frac{\cos\theta}{M} \delta(x^1)\right) \tan\theta \right\}_2 \tag{7.23b}$$

or simplifying

$$S_{12}^{\cdot\cdot 1} = \left\{ -\left(\tfrac{1}{5}\right) \delta(x^1) \right\}_1 + \left\{ -\left(\tfrac{1}{5}\right) \delta(x^1) \right\}_2 \tag{7.23c}$$

where the term $(\cos\theta / M\delta(x^1))$ in Eq. 7.23b follows from the fact that

$$\frac{\partial H(x^{k^{I3}})}{\partial x^{k^{I3}}} = \frac{\partial H(x^{k^{I3}})}{\partial x^{k^{I3c}}} \frac{\partial x^{k^{I3c}}}{\partial x^{k^{I3}}} \tag{7.24a}$$

or

$$\delta(x^{k^{I3}}) = (x^{k^{I3c}}) C_{k^{I3}}^{k^{I3c}} \tag{7.24b}$$

which in view of Eq. 7.13, becomes

$$\delta(x^1) = \delta(x^1)\frac{\cos^2\theta}{M} \tag{7.24c}$$

The above arguments follow along the same lines that led to Eq. 4.108. When Eq. 7.23c is substituted into the following relation

$$b^{k^{13c}} = -\int_S S_{l^{13c}m^{13c}}{}^{k^{13c}}\, dF^{l^{13c}m^{13c}} \tag{7.25}$$

we obtain the same result as that given by Eqs. 7.19.

It is also a simple matter to connect the (κ^I) and (k') states of Figures 6.8 and 4.2, respectively, in terms of a simple coordinate transformation $C_{k'}^{\kappa^I}$ that gives

$$A_K^{\kappa^I} = C_{k'}^{\kappa^I} A_K^{k'} = \delta_K^{\kappa^I} \tag{7.26}$$

where $A_K^{k'}$ is given by Eq. 4.16, while

$$C_{k'}^{\kappa^I} = \left\{\delta_{k'}^K \delta_\lambda^{\kappa^I} B_K^\lambda H(-x^1)\right\}_1 + \left\{\delta_{k'}^{\kappa^I} H(+x^1)\right\}_2 \tag{7.27}$$

Similarly, the elastic (κ^I) state can be converted to the plastic (k') state as follows:

$$A_K^{k'} = C_{\kappa^I}^{k'} A_K^{\kappa^I} = C_{\kappa^I}^{k'} \delta_K^{\kappa^I} \tag{7.28}$$

whereas

$$B_{k'}^K = C_{k'}^{\kappa^I} B_{\kappa^I}^K = \delta_{k'}^K \tag{7.29}$$

It should be pointed out that the coordinate transformations discussed above are rather special in that they convert a plastic strain into an elastic strain or vice versa. They would not be used to represent an undeformed state in terms of dislocations, since according to Eqs. 7.28 and 7.29, we need to write for this case

$$A_K^{k'} = C_{\kappa^I}^{k'} \delta_K^{\kappa^I} \tag{7.30a}$$

and

$$B_{k'}^K = C_{k'}^{\kappa^I} \delta_{\kappa^I}^K \tag{7.30b}$$

The last equation violates the principle that in order to have plastic distortion, the metric must be dragged. Note, however, that Eq. 7.29 satisfies this requirement. Thus, the interphase distortion associated with the (κ^I) state shown in Figure 6.8*a* can be described either in terms of virtual dislocations using the $Q_{\mu'\lambda'}^{\cdot\cdot\kappa^I}$ tensor, as in Eq. 6.108, or in terms of dislocations using the coordinate transformation given by Eq. 7.27.

It is important to point out that although the net Burgers vectors associated with the (k^{I3c}) and (k^{I10c}) states of Figures 7.1*a* and 7.2, respectively, are zero, there are nevertheless dislocations at the boundary. This is most easily seen by reference to the latter figure where such dislocations are revealed in terms of the coincidence site lattice sublattice. Since they occur in equal number, but opposite sign within each coincidence site lattice, they cancel with one another and so are not revealed in such a lattice. The same occurs for the (k^{I3c}) state of Figure 7.1*a* which is redrawn in Figure 7.4*a* in terms of the coincidence site lattice sublattice. Each coincidence site lattice unit cell is seen to consist of an array of dislocation dipoles within the boundary so that the net Burgers vector associated with this array within such a lattice is again zero. Figure 7.4*b* is simply the relaxed state of Figure 7.4*a* after the two grains are separated by tearing at the boundary. Also important to note here is that the two tilt boundaries illustrated in Figures 6.1*a* and 6.1*b* represent extreme cases. In the former configuration, there is no free surface at the boundary but the elastic distortion is a maximum. In the latter case, there is no elastic distortion at the boundary, but the amount of free surface is a maximum. In general, there is a compromise between these two extremes such as shown in Figure 7.5. More is said about this important consideration when the energy of grain boundaries is considered in another section.

It has often been argued that there can be no such thing as a 90° grain boundary. Formally such a statement is erroneous. This can be seen by referring to Figure 7.6*a* which is the 90° torn counterpart of the 53.1° boundary shown in Figure 6.1*b*. The open circles represent the positions of the atoms so that the link between the continuum and discrete atom representations of a grain boundary become somewhat more clear. It also follows that the free surfaces associated with Figure 7.6*a* could be eliminated by the introduction of large strains in the manner employed to obtain Figure 6.1*a*. Such distortions, however, would be severe and an energetically more favorable manner in which these surfaces could be eliminated would be by the introduction of atoms in the way depicted in Figure 7.6*b*. Such a boundary contains dislocations, that is, possesses a nonvanishing torsion tensor, but is unique in that it displays neither free surfaces nor elastic distortions.

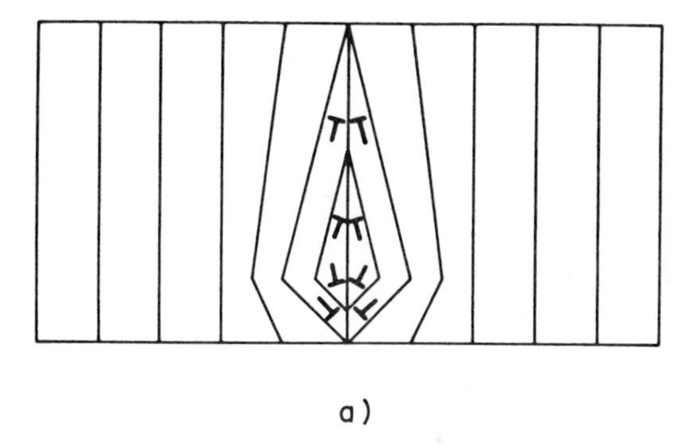

a)

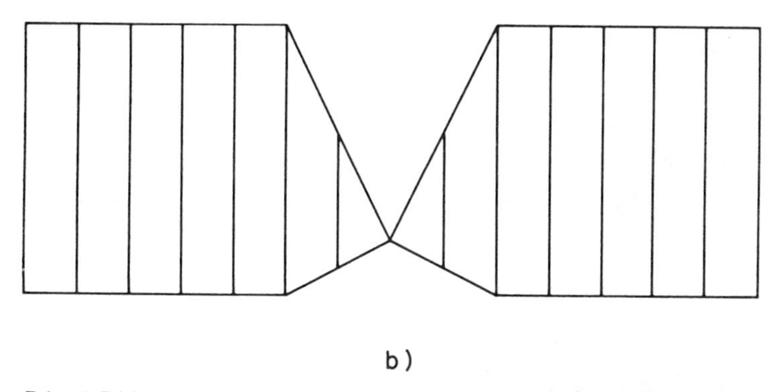

b)

Figure 7.4 *a*) Dislocation content of the grain boundary shown in Figure 6.1*d* described in terms of the coincidence site lattice sublattice. *b*) Torn configuration of *a*).

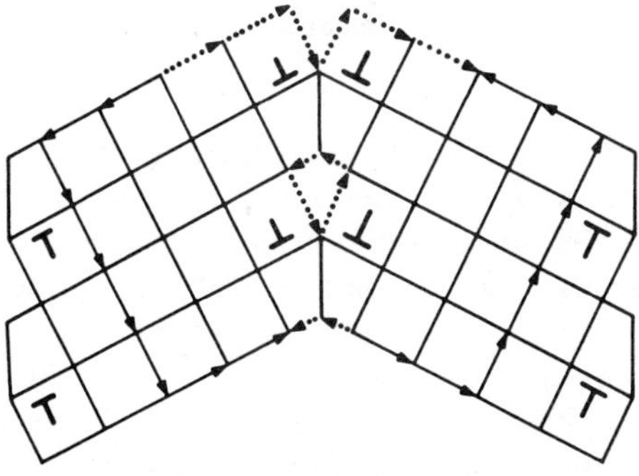

Figure 7.5 Intermediate configuration of a symmetric tilt boundary between the two extreme cases shown in Figure 6.1*a* and 6.1*b*.

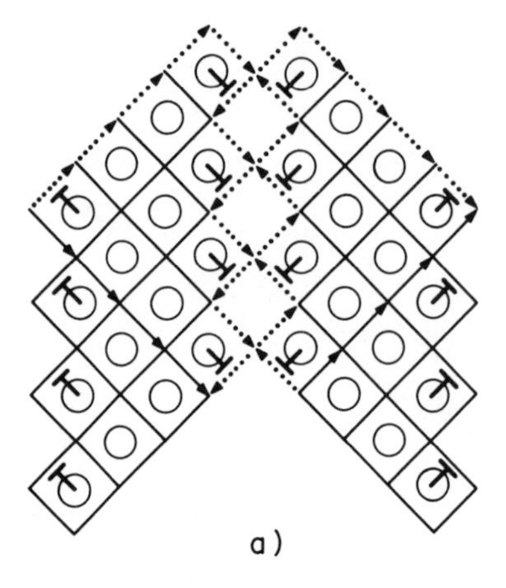

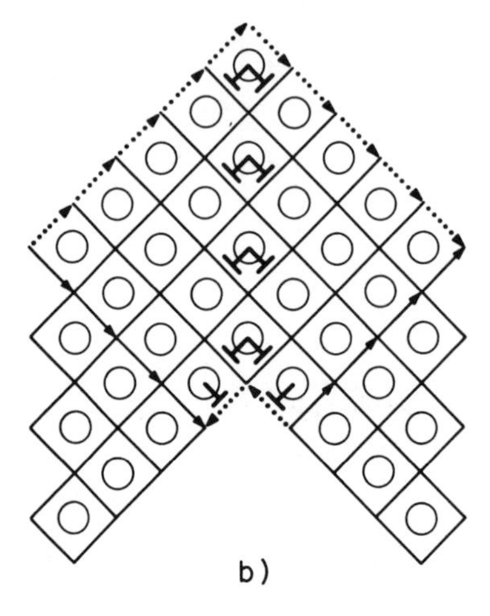

Figure 7.6 *a*) Torn configuration associated with a symmetric 90° tilt boundary. *b*) Same configuration as in *a*), but after removal of free surfaces at the boundary.

Finally, in order to demonstrate the generality of the present formulation, Figure 7.7 shows the representation of the coherent boundary of Figure 6.10*a* in terms of a common or coincidence site lattice. The dislocation content of such a boundary is readily apparent and the relevant tensor quantities associated with this boundary can be easily obtained. Figure 7.7 can also be compared with Figure 5.24. If the dislocation array associated with the rightmost portion of the crystal shown in Figure 5.24*a* were incorporated into Figure 7.7, then the interface would generate no long-range stresses in much the same way as the boundaries shown in

Figures 6.9c and 6.1d, and a perfect coincidence site lattice could be associated with this configuration. Also important to note is that whereas the dislocation array shown in Figure 7.7 is related to the strains associated with the phase transformation, those shown in Figure 5.24a are related to the strains associated with the externally applied stress. Thus we are now in a position to argue that it makes no difference how a distortion arises within a body. In any case, a common or coincidence site lattice can be constructed so as to give a dislocation representation of these distortions within that body. It is apparent that the coincidence site lattice provides a common metric with which to describe the distortion of a body. In particular, the two phases shown in Figures 6.8a and 6.10a, except for the elastic distortions, are given a common metric. In the case of Figure 7.2, this is simply given by

$$g_{k^{I10c}l^{I10c}} = \mathbf{e}_{k^{I10c}} \cdot \mathbf{e}_{l^{I10c}} \tag{7.31}$$

where $\mathbf{e}_{k^{I10c}}$ are the base vectors which coincide with the edges of the coincidence site lattice. Thus we arrive at the very powerful concept that any internal elastic distortion can be represented in terms of a nonvanishing $Q_{ml}^{\cdot\cdot\,k}$ tensor that allows the metric to vary with position, or in terms of a common lattice or metric that allows these distortions to be represented in terms of dislocations, that is, a nonvanishing $S_{ml}^{\cdot\cdot\,k}$.

The (k^{I10}) state of Figure 6.9c, as considered thus far, has been assumed to be metric, that is, $Q_{m^{I10}l^{I10}}^{\cdot\cdot\,k^{I10}} = 0$. However, similar to Eq. 6.111 it could be made nonmetric by writing

$$Q_{m^{I10}l^{I10}}^{\cdot\cdot\,k^{I10}} = \delta_{m^{I10}}^{\mu^{I}} \delta_{l^{I10}}^{\lambda^{I}} \delta_{\kappa^{I}}^{k^{I10}} Q_{\mu^{I}\lambda^{I}}^{\cdot\cdot\,\kappa^{I}} \tag{7.32}$$

Since the (k^{I10}) state has already been shown to possess a torsion tensor

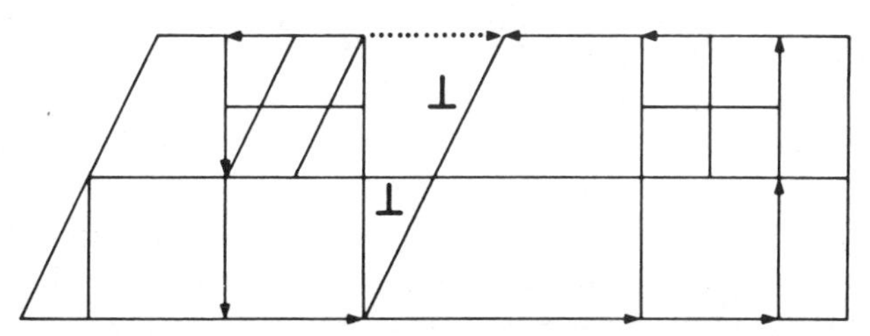

Figure 7.7 Representation of the coherent boundary shown in Figure 6.10a in terms of dislocations.

given by Eq. 6.82, it follows that

$$Q_{m^{I10}l^{I10}}^{\cdot\ \cdot\ k^{I10}} = -2S_{m^{I10}l^{I10}}^{\cdot\ \cdot\ k^{I10}} \tag{7.33}$$

so that

$$b^{k^{I10}} = -\int_s \left[\frac{1}{2} Q_{m^{I10}l^{I10}}^{\cdot\ \cdot\ k^{I10}} + S_{m^{I10}l^{I10}}^{\cdot\ \cdot\ k^{I10}} \right] dF^{m^{I10}l^{I10}} = 0 \tag{7.34}$$

In physical terms, this means that the real dislocations corresponding to $S_{m^{I10}l^{I10}}^{\cdot\ \cdot\ k^{I10}}$ just cancel with the virtual dislocations associated with $Q_{m^{I10}l^{I10}}^{\cdot\ \cdot\ k^{I10}}$. It also now becomes clear that the (κa) state of Figure 5.16 that was generated by an externally applied stress, except for some localized elastic strains at the boundary, is identical to the (k^{I10}) state of Figure 6.9*c*. There at first however seems to be a paradox, since Figure 6.9*c* possesses no long-range stress, while Figure 5.16*b* has associated with it internal strains. However, these internal stresses are just balanced by the external stress. Once again, the equivalence between an externally and internally stressed body is demonstrated.

REVIEW

Perhaps the best and simplest way in which to understand the concept of a common or coincidence site lattice is in terms of the (k') and (κ^I) states shown in Figures 4.2 and 6.8*a*, respectively. It is apparent that the distortions in both cases are similar; however, in the former case the state is elastic whereas in the latter case we are dealing with a dislocated state. These states can, however, be related to one another by a coordinate transformation $C_{k'}^{\kappa^I}$ given by Eq. 7.27. It is important to note that this transformation is not a mathematically admissible transformation since, as the equation shows, it is not a continuous function, but rather is in the form of a Heaviside function. These transformations are thus rather special in that they convert an elastic distortion into a plastic one and vice versa. It is clear from Figures 4.2 and 6.8*a* that in going from state (κ^I) to state (k') that a common lattice or metric is created, except of course for the elastic strains in the vicinity of the dislocation. Thus it is this concept associated with the creation of a common or coincidence site lattice that allows elastic distortions to be described in terms of dislocations. The transformation can also be reversed such that a dislocated state is expressed in terms of an elastically strained state.

CHAPTER 8

Small Distortions and the Linear Approximation

The treatment given to elastic and plastic distortions in Chapters 2 and 3, respectively, has been quite general and is therefore applicable to arbitrarily large distortions. Most of the studies to date, however, dealing with this subject have restricted themselves to small distortions. Under these conditions, it is possible to linearize all of the relations in turn leading to greatly simplified expressions for the distortions. In particular, it is shown that the difference between covariant and contravariant indices is lost so that all of the relevant tensor quantities can be expressed solely in terms of covariant indices.

Thus far, our analysis of distortions, both elastic as well as plastic, has been most general. On the other hand, nearly all of the development of this subject has been based upon small distortions, that is, the linearized approximation. The reason for this is obviously due to the fact that the constituitive relations between stress and strain are based upon Hooke's law that is in fact a linear relation. It is extremely important to understand the linearization procedures carried out with respect to the more generalized relations, since much is lost by such a process in exchange for simplification. Let us begin by once again considering the distorted state shown in Figure 2.2*a*. This time, however, we consider the body to be imbedded in a rectangular Cartesian coordinate frame such as illustrated in Figure 8.1. Under these conditions, the metric tensors given by Eqs. 2.2 and 2.6 now become

$$g_{\kappa\lambda} = a_{KL}\delta_\kappa^K\delta_\lambda^L = \delta_{\kappa\lambda} \tag{8.1}$$

We may also define a displacement vector $\mathbf{u}$ such as illustrated in Figure

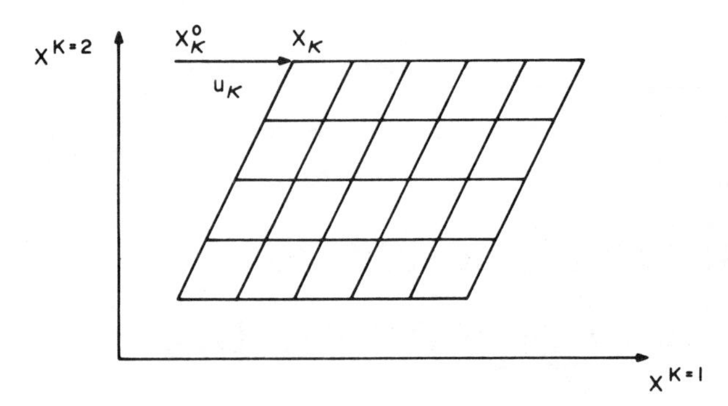

Figure 8.1 Elastic distortions shown in Figure 2.2*a* but imbedded in a rectangular Cartesian coordinate system.

8.1 with components given by

$$u^K = x_f^K - x^K \tag{8.2}$$

where x^K and x_f^K correspond to the initial and final coordinates measured with respect to the rigid Cartesian coordinate system. We could also write

$$x_f^K = \delta_\kappa^K x^\kappa \tag{8.3}$$

that allows Eq. 8.2 to be written as

$$u^K = \delta_\kappa^K x^\kappa - x^K \tag{8.4}$$

Upon differentiation of the last expression with respect to x^κ, we obtain

$$\frac{\partial x^K}{\partial x^\lambda} = \delta_\lambda^K - \frac{\partial u^K}{\partial x^\lambda} = A_\lambda^K \tag{8.5}$$

that when substituted into Eq. 2.17 yields (Fung, 1965)

$$e_{\kappa\lambda} = \frac{1}{2}\left(\delta_{\kappa\lambda} - \delta_{KL}\frac{\partial x^K}{\partial x^\kappa}\frac{\partial x^L}{\partial x^\lambda}\right) \tag{8.6}$$

or

$$e_{\kappa\lambda} = \frac{1}{2}\left(\frac{\partial u^\kappa}{\partial x^\lambda} + \frac{\partial u^\lambda}{\partial x^\kappa} - \frac{\partial u^K}{\partial x^\kappa}\frac{\partial u^K}{\partial x^\lambda}\right) \tag{8.7}$$

In the linear approximation, we can neglect the higher order terms and obtain the well-known expression for the strain given by (Nye, 1957)

$$e_{\kappa\lambda} = \frac{1}{2}\left(\frac{\partial u_\kappa}{\partial x_\lambda} + \frac{\partial u_\lambda}{\partial x_\kappa}\right) \tag{8.8}$$

where the subscripts have been lowered since in orthogonal Cartesian coordinates, the distinction between covariance and contravariance vanishes. In this respect, Eq. 8.2 could also be simplified to read

$$u_\kappa = x_\kappa - x_\kappa^o \tag{8.9}$$

where x_κ and x_κ^o are the final and initial coordinates measured in the Cartesian system. It follows therefore that in the limit of $\theta \to 0$, the strain tensor associated with the (κ^2) state of Figure 2.2*a* is given by

$$e_{\kappa^{2'}\lambda^{2'}} = \begin{bmatrix} 0 & \frac{1}{2}\theta & 0 \\ \frac{1}{2}\theta & 0 & 0 \\ 0 & 0 & 0 \end{bmatrix} \tag{8.10}$$

where primes are used to denote linearization. This is precisely the result that Eq. 2.26 reduces to as $\theta \to 0$.

It is instructive at this point to give a somewhat more formal proof showing why the distinction between contravariance and covariance disappears in rectangular Cartesian coordinates (Spain, 1953). In the first place, the conditions for rectangular Cartesian coordinates is that

$$(ds)^2 = dx^K dx^K = A_\kappa^K A_\lambda^K dx^\kappa dx^\lambda = \delta_{\kappa\lambda} dx^\kappa dx^\lambda \tag{8.11}$$

so that

$$A_\kappa^K A_\lambda^K = \delta_{\kappa\lambda} \tag{8.12}$$

The last equation obviously does not hold for arbitrarily large distortions as can be seen by considering the (κ^4) state of Figure 2.3. In this particular case

$$A_{\kappa^4}^K = \delta_{\kappa^4}^{a^4} A_{a^4}^K \tag{8.13}$$

where $A_{a^4}^{K}$ can be obtained from Eq. 3.20. In particular

$$A_{\kappa^4}^{K}=\begin{bmatrix} 1 & \tan\theta & 0 \\ -\tan\theta & 1 & 0 \\ 0 & 0 & 1 \end{bmatrix} \equiv \begin{bmatrix} 1 & \theta & 0 \\ -\theta & 1 & 0 \\ 0 & 0 & 1 \end{bmatrix} = \begin{bmatrix} A_{11} & A_{12} & A_{13} \\ A_{21} & A_{22} & A_{23} \\ A_{31} & A_{32} & A_{33} \end{bmatrix} = A_{\lambda^4\kappa^4} \tag{8.14a}$$

while

$$A_{K}^{\kappa^4}=\begin{bmatrix} \cos^2\theta & -\sin\theta\cos\theta & 0 \\ \sin\theta\cos\theta & \cos^2\theta & 0 \\ 0 & 0 & 1 \end{bmatrix} \equiv \begin{bmatrix} 1 & -\theta & 0 \\ \theta & 1 & 0 \\ 0 & 0 & 1 \end{bmatrix} = A_{\kappa^4\lambda^4} \tag{8.14b}$$

The first set of matrices in the last two equations obviously do not satisfy Eq. 8.12, whereas the last two, which are the linear approximations, do. It also follows that in linearized form

$$A_{K}^{\kappa^4}=A_{\kappa^4}^{K} \tag{8.15}$$

which means that the indices K and κ^4 loose their significance so that we may lower them to read

$$A_{K\kappa^4}=A_{\kappa^4K}=A_{\kappa^4\lambda^4} \tag{8.16}$$

This also means that Eq. 8.12 can be rewritten as

$$A_{\mu\kappa}A_{\mu\lambda}=\delta_{\kappa\lambda} \tag{8.17}$$

whereas the coordinates can be related as follows:

$$dx'_{\kappa}=A_{\kappa\lambda}\,dx_{\lambda} \tag{8.18a}$$

and

$$dx_{\kappa}=A_{\lambda\kappa}\,dx'_{\lambda} \tag{8.18b}$$

where the primed quantities refer to the "new" or distorted system, whereas the unprimed quantities refer to the "old" or undistorted reference system.

Now the quantities $\partial u_{\kappa}/\partial x_{\lambda}$ that appear in Eq. 8.8 may be referred to as elastic distortions. They are obviously tensors and may be written as

(Kröner, 1958, 1966; deWit, 1960)

$$\beta_{\kappa\lambda} = \frac{\partial u_\kappa}{\partial x_\lambda} \tag{8.19}$$

Since $\beta_{\kappa\lambda}$ can be expressed as the sum of a symmetric and asymmetric part, we may write

$$\beta_{\kappa\lambda} = e_{\kappa\lambda} + \omega_{\kappa\lambda} \tag{8.20}$$

where

$$e_{\kappa\lambda} = \tfrac{1}{2}(\beta_{\kappa\lambda} + \beta_{\lambda\kappa}) = e_{\lambda\kappa} \tag{8.21a}$$

which is just the strain tensor given by Eq. 8.8, while the antisymmetric tensor $\omega_{\kappa\lambda}$ is given by

$$\omega_{\kappa\lambda} = -\frac{1}{2}(\beta_{\kappa\lambda} - \beta_{\lambda\kappa}) = -\omega_{\lambda\kappa} \tag{8.21b}$$

For the linearized version of the (κ^2) state shown in Figure 2.2*a*, the only nonvanishing component of $\beta_{\kappa\lambda}$ is given by

$$\beta_{12} = \theta \tag{8.22}$$

The distortions can be used to determine the displacements as follows:

$$du_\kappa = \beta_{\kappa\lambda}\, dx_\lambda \tag{8.23}$$

At this point it is important to note that the distortions given by $\beta_{\kappa\lambda}$ are fundamentally different from those given by A_K^κ in Eq. 2.13. The former tensors deal with displacements, while the latter are related to coordinate transformations. Equation 8.3, however, may be used to relate these two distortions as follows:

$$\delta_\lambda^K - \beta_\lambda^K = A_\lambda^K \tag{8.24}$$

At this point we arrive at a very important concept and that is the fact that it was not Eq. 8.1 alone that allowed us to dispense with the difference between covariant and contravariant indices. In addition, it was also necessary to linearize the distortions. The use of Eq. 8.1 alone was in fact used to define a corresponding plastic strain as can be seen by reference to Eq. 3.13. The implications behind the linearization process can be seen by considering the strain tensor associated with the (κ^6) state of Figure 2.4 and given by Eq. 2.40. It is seen that in the limit as $\theta \to 0$, $e_{\kappa^6\lambda^6}$, except for a factor of two, becomes identical to that given by $e_{\kappa^{2'}\lambda^{2'}}$ in Eq. 8.10. This is simply related to the fact that in the linear approximation, a simple shear is equivalent to a pure shear plus a rigid rotation, that is (Nye, 1957),

$$\beta_{\kappa^{2'}\lambda^{2'}} = \beta^{e}_{\kappa^{2'}\lambda^{2'}} + \beta^{\omega}_{\kappa^{2'}\lambda^{2'}} \tag{8.25a}$$

or in matrix form

$$\begin{bmatrix} 0 & 0 & 0 \\ \theta & 0 & 0 \\ 0 & 0 & 0 \end{bmatrix} = \begin{bmatrix} 0 & \frac{\theta}{2} & 0 \\ \frac{\theta}{2} & 0 & 0 \\ 0 & 0 & 0 \end{bmatrix} + \begin{bmatrix} 0 & -\frac{\theta}{2} & 0 \\ \frac{\theta}{2} & 0 & 0 \\ 0 & 0 & 0 \end{bmatrix} \tag{8.25b}$$

where it is also clear that

$$\beta_{\kappa^{6'}\lambda^{6'}} = 2\beta^{e}_{\kappa^{2'}\lambda^{2'}} \tag{8.25c}$$

Turning now to the linearized form of state (κ^4), it is clear from Eq. 2.32 that the strain is zero and is due to the fact that under these conditions, a rigid rotation obtains. It follows then that we can write

$$\beta_{\kappa^{4'}\lambda^{4'}} = \beta_{\kappa^{3'}\lambda^{3'}} = \begin{bmatrix} 0 & -\theta & 0 \\ \theta & 0 & 0 \\ 0 & 0 & 0 \end{bmatrix} \tag{8.26}$$

It is obvious from Eq. 8.21a that the strain is

$$e_{\kappa^{4'}\lambda^{4'}} = e_{\kappa^{3'}\lambda^{3'}} = 0 \tag{8.27}$$

while the rotation given by Eq. 8.21b is

$$\omega_{\kappa^{4'}\lambda^{4'}} = \omega_{\kappa^{3'}\lambda^{3'}} = \begin{bmatrix} 0 & \theta & 0 \\ -\theta & 0 & 0 \\ 0 & 0 & 0 \end{bmatrix} \tag{8.28}$$

It is apparent that in the previous two cases

$$\omega_{\kappa^{2'}\lambda^{2'}} = \omega_{\kappa^{6'}\lambda^{6'}} = 0 \tag{8.29}$$

It is also clear that in the linearized approximation, Eq. 8.24 can be written as

$$\delta_{\kappa\lambda} - \beta_{\kappa\lambda} = A_{\kappa\lambda} \tag{8.30}$$

where in Eq. 8.24

$$A_{\lambda}^{K} = \delta_{\kappa}^{K} \delta_{\lambda}^{L} B_{L}^{\kappa} \tag{8.31}$$

which is analogous to the relation given by Eq. 3.9a for plastic distortions. Thus for the $(\kappa^{4'})$ state, we obtain from Eq. 8.30

$$A_{12} = -\beta_{12} = \theta \tag{8.32a}$$

and

$$A_{21} = -\beta_{21} = -\theta \tag{8.32b}$$

while

$$A_{11} = A_{22} = A_{33} = 1 \tag{8.32c}$$

which is in agreement with the results obtained from Eq. 8.31, when used in conjunction with the linearized form of Eq. 2.30. We thus see that the distortions given by $A_{\kappa\lambda}$ are related to coordinate transformations, while those given by $\beta_{\kappa\lambda}$ are related to displacements. It also follows that for the linearized forms of the (κ), (κ^1), and (κ^5) states, we have nonvanishing components e_{22} for all three, with an additional component e_{22} for the last. The rotations $\omega_{\kappa\lambda}$ vanish in all three cases.

The plastic distortion such as shown in Figure 3.2*b* can also be imbedded in a rectangular Cartesian coordinate frame, as illustrated in Figure 8.2*a*. Under these conditions, we may write

$$g_{ab} = a_{KL} \delta_{a}^{K} \delta_{b}^{L} = \delta_{ab} \tag{8.33}$$

However, this is precisely the procedure that has already been carried out to describe plastic distortion in Section 3. We now proceed to analyze these distortions in the limit of small strains. Similar to Eq. 8.2, we can define a displacement vector with components

$$u^{a} = x^{a} - x_{o}^{a} \tag{8.34}$$

In addition

$$x_{o}^{a} = \delta_{K}^{a} x^{K} \tag{8.35}$$

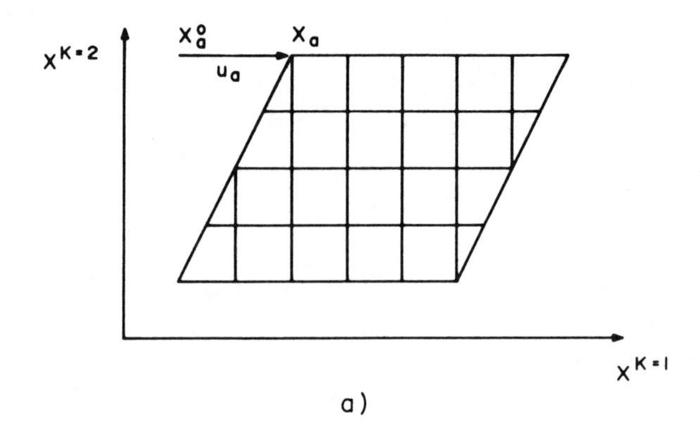

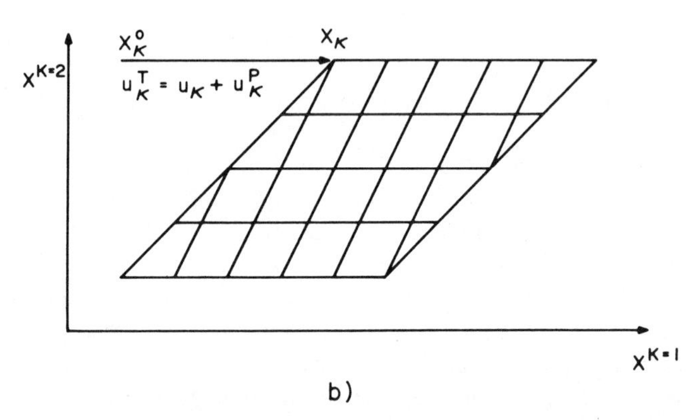

Figure 8.2 *a*) Plastic distortion shown in Figure 3.2*b* but imbedded in a rectangular Cartesian coordinate system. *b*) Sum of elastic plus plastic distortions shown in Figures 8.1 and 8.2*a*.

so that Eq. 8.34 becomes

$$u^a = x^a - \delta_K^a x^K \tag{8.36}$$

which upon differentiation becomes

$$\frac{\partial x^a}{\partial x^L} = \frac{\partial u^a}{\partial x^L} + \delta_L^a = A_L^a \tag{8.37}$$

This can be substituted into Eq. 3.13 to obtain the plastic strain as follows:

$$e_{KL}^P = \frac{1}{2}\left(\delta_{ab}\frac{\partial x^a}{\partial x^K}\frac{\partial x^b}{\partial x^L} - \delta_{KL}\right) \tag{8.38}$$

or

$$e_{KL}^{P}=\frac{1}{2}\left(\frac{\partial u^{K}}{\partial x^{L}}+\frac{\partial u^{L}}{\partial x^{K}}+\frac{\partial u^{a}}{\partial x^{K}}\frac{\partial u^{a}}{\partial x^{L}}\right) \tag{8.39}$$

or in linearized form

$$e_{ab}^{P}=\frac{1}{2}\left(\frac{\partial u_{a}}{\partial x_{b}}+\frac{\partial u_{a}}{\partial x_{b}}\right) \tag{8.40}$$

where Eq. 8.34 can now be written as

$$u_{a}=x_{a}-x_{a}^{o} \tag{8.41}$$

We thus see that the plastic strains given by Eq. 8.40 become formally identical to the elastic strains given by Eq. 8.8. In addition, the plastic distortions can be written as

$$\beta_{ab}^{P}=\frac{\partial u_{a}}{\partial x_{b}} \tag{8.42}$$

where again, similar to Eq. 8.20, we can write

$$\beta_{ab}^{P}=e_{ab}^{P}+\omega_{ab}^{P} \tag{8.43}$$

as well as

$$du_{a}=\beta_{ab}^{P}dx_{b} \tag{8.44}$$

In the more generalized case where both elastic as well as plastic distortion occur, we may write (Kröner, 1958)

$$u_{\kappa}^{T}=u_{\kappa}+u_{\kappa}^{P} \tag{8.45}$$

where u_{κ}^{T} is the total displacement. Thus,

$$du_{\kappa}^{T}=du_{\kappa}+du_{\kappa}^{P} \tag{8.46}$$

so that we may write

$$\beta_{\kappa\lambda}^{T}=\beta_{\kappa\lambda}+\beta_{\kappa\lambda}^{P} \tag{8.47}$$

where $\beta_{\kappa\lambda}^{T}$ is the tensor of total distortion. It also follows that

$$du_{\kappa}^{T}=\beta_{\kappa\lambda}^{T}dx_{\lambda} \tag{8.48}$$

Similar to Eq. 8.20 the total plastic distortion can be written as the sum of a pure strain and a rotation, that is,

$$\beta_{\kappa\lambda}^{P} = e_{\kappa\lambda}^{P} + \omega_{\kappa\lambda}^{P} \tag{8.49}$$

It also follows that the total strain is given by

$$e_{\kappa\lambda}^{T} = e_{\kappa\lambda} + e_{\kappa\lambda}^{P} \tag{8.50}$$

The elastic strain for the linear approximation could also have been given by Eq. 2.18 rather than Eq. 8.6; that is, by

$$e_{\kappa^{2'}\lambda^{2'}} = \tfrac{1}{2}(g_{\kappa^{2'}\lambda^{2'}} - \delta_{\kappa^{2'}\lambda^{2'}}) \tag{8.51}$$

In this case, the linearized form of the metric tensor given by Eq. 2.25 would be required, or

$$g_{\kappa^{2'}\lambda^{2'}} = \begin{bmatrix} 1 & \theta & 0 \\ \theta & 1 & 0 \\ 0 & 0 & 1 \end{bmatrix} \tag{8.52}$$

and would lead to the same expression for strain as that given by Eq. 8.10. Thus, we have two equivalent definitions, one given by Eq. 8.8 based upon a dragged metric tensor, and the other given by Eq. 2.18 based upon dragged coordinates. Equating them gives

$$2e_{\kappa\lambda} = g_{\kappa\lambda} - \delta_{\kappa\lambda} = \frac{\partial u_\kappa}{\partial x_\lambda} + \frac{\partial u_\lambda}{\partial x_\kappa} \tag{8.53}$$

Another interesting outcome of Eq. 8.53 can be seen by first writing the compatability conditions for elastic strain as follows (Sokolnikoff, 1964; deWit, 1978)

$$4(\partial_{\mu\rho} e_{\kappa\lambda})_{[\mu\lambda][\rho\kappa]} \equiv \partial_{\mu\rho} e_{\kappa\lambda} - \partial_{\lambda\rho} e_{\kappa\mu} + \partial_{\lambda\kappa} e_{\rho\mu} - \partial_{\mu\kappa} e_{\rho\lambda} = 0 \tag{8.54}$$

where the square brackets denote that only the antisymmetric components are to be considered. It is also possible to write an expression for a quantity termed the Riemann–Christoffel curvature tensor of the first kind given by

$$\begin{aligned} R_{\mu\lambda\rho\kappa} = {} & \tfrac{1}{2}(\partial_{\mu\rho} g_{\kappa\lambda} - \partial_{\lambda\rho} g_{\kappa\mu} + \partial_{\lambda\kappa} g_{\rho\mu} - \partial_{\mu\kappa} g_{\rho\lambda}) \\ & + g^{\nu\gamma}([\mu\nu\kappa][\lambda\rho\gamma] - [\lambda\nu\kappa][\mu\rho\gamma]) \end{aligned} \tag{8.55}$$

where $[\mu\nu\kappa]$ are Christoffel symbols of the first kind given by (Sokolnikoff, 1964)

$$[\mu\nu\kappa] = \tfrac{1}{2}(\partial_\mu g_{\nu\kappa} + \partial_\nu g_{\mu\kappa} - \partial_\kappa g_{\mu\nu}) \tag{8.56}$$

In the linearized form, Eq. 8.55 may be written as

$$R_{\mu\lambda\rho\kappa} \equiv 2(\partial_{\mu\rho} g_{\kappa\lambda})_{[\mu\lambda][\rho\kappa]} = \tfrac{1}{2}(\partial_{\mu\rho} g_{\kappa\lambda} - \partial_{\lambda\rho} g_{\kappa\mu} + \partial_{\lambda\kappa} g_{\rho\mu} - \partial_{\mu\kappa} g_{\rho\lambda}) \tag{8.57}$$

so that from Eq. 8.53, along with Eqs. 8.54 and 8.57 we have

$$4(\partial_{\mu\rho} e_{\kappa\lambda})_{[\mu\lambda][\rho\kappa]} = R_{\mu\lambda\rho\kappa} \tag{8.58}$$

so that in the linear approximation, the vanishing of the curvature of the first kind is synonymous with satisfying the compatability conditions.

Now just as it was possible to imbed an oblique elastic distortion in a rectangular Cartesian coordinate system, it is also possible to imbed a rectangular Cartesian plastic distortion such as shown in Figure 8.2*a* in an oblique coordinate system, as illustrated in Figure 8.3. This means that the plastic strains given by Eq. 3.13 can now be written as

$$e^P_{ab} = \tfrac{1}{2}(\delta_{KL} C^K_a C^L_b - \delta_{ab}) = \tfrac{1}{2}(g_{ab} - \delta_{ab}) \tag{8.59}$$

where the C^K_a are coordinate transformations given by

$$C^K_a = \delta^b_a \delta^K_L A^L_b \tag{8.60}$$

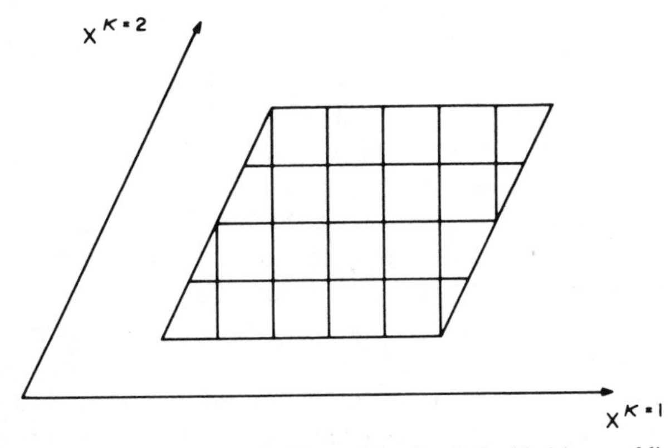

Figure 8.3 Plastic distortions shown in Figure 8.2*a*, but imbedded in an oblique coordinate system.

where the A_b^L are plastic distortions as defined in Section 3. We thus have

$$g_{ab} = C_a^K C_b^L \delta_{KL} \tag{8.61}$$

so that the g_{ab} is the metric δ_{KL} transformed into another coordinate system.

Thus it now becomes possible to write an expression similar to Eq. 8.53 for the plastic strain or

$$2e_{ab}^P = g_{ab} - \delta_{ab} = \frac{\partial u_a}{\partial x_b} + \frac{\partial u_b}{\partial x_a} \tag{8.62}$$

We could also write a compatability condition for plastic strain similar to that given by Eq. 8.54 and a curvature tensor similar to that given by Eq. 8.55 so as to finally obtain

$$4(\partial_{cd} e_{ab}^P)_{[cb][da]} = R_{cbda}^P \tag{8.63}$$

which is similar to Eq. 8.58 for elastic strain.

REVIEW

Many problems dealing with the mechanical behavior of solids involve small distortions. Under these conditions, the same Cartesian coordinate system may be used to describe both the deformed and undeformed bodies. The tensor quantities associated with such distortions in turn become greatly simplified. In particular, we are able to lower all of the indices into covariant positions and redefine a somewhat modified elastic distortion tensor $\beta_{\kappa\lambda}$ in the form given by Eq. 8.24. Also, in contrast to an arbitrarily large distortion, we are able to separate $\beta_{\kappa\lambda}$ into a pure strain and a pure rotation as expressed in Eq. 8.20. This can be seen more clearly by reconsidering the simply sheared (κ^2) state of Figure 2.2*a* in the linear approximation. Under these conditions, the distortion tensor can be decomposed into a pure shear plus a rigid rotation as embodied in Eq. 8.25. This was not possible for an arbitrarily large single shear. Similarly, a plastic distortion tensor β_{ab}^P can also be formulated for small distortions that can be treated in a manner analagous to that for elastic distortions. All of the tensor quantities such as the distortion, strain, and metric associated with the elastic and plastic distortions contained in Chapters 2 and 3, respectively, have been reformulated in terms of the linear approximation in the present chapter. These quantities in turn serve as the foundations of the linearized formulations of dislocation theory to be considered in the following chapter.

CHAPTER 9

Dislocations in the Linear Approximation

Similar to what was done in the previous chapter for distortions, the tensor quantities associated with dislocations, such as were introduced in Chapters 4 and 5, can also be linearized. This linearization allows Cartesian coordinates to be maintained within both the dislocated and undislocated states and leads to immense simplification of such tensor quantities as torsion and dislocation density. Again, as in the previous chapter, it is primarily within the confines of the linear approximation that most of the present-day theory of dislocations concerns itself.

Similar to Eq. 8.24, Eq. 8.36 can be differentiated to obtain the following relationship for plastic distortion:

$$\delta_L^a + \beta_L^a = A_L^a \tag{9.1a}$$

or in the linear approximation

$$\delta_{ab} + \beta_{ab}^P = A_{ab} \tag{9.1b}$$

Let us now consider the linearized case of the (k') state of Figure 4.2. According to Eqs. 9.1 and 4.37, we can write (Kröner, 1964)

$$b_{k'} = -\oint A_{k'l'}\,dx_{l'} = -\oint \beta_{k'l'}^P\,dx_{l'} \tag{9.2}$$

According to Eq. 4.11, on the other hand, the only nonvanishing component of the distortion is given by

$$\beta_{22}^P = \{(V-1)H(-x_1)\}_1 \tag{9.3}$$

where $(V-1)$ is a positive number much less the unity. Combining the last two equations, we obtain

$$b_{k'=2}=-\int_3^4 dx_2=4(V-1) \tag{9.4}$$

This is similar to the result given by Eq. 4.42, except that the lengths of the dotted closure failure segments in Figure 4.2 are vanishingly small. The line integral of Eq. 9.2 could also be converted into a surface integral by means of Stokes' theorem, similar to Eq. 4.58, to obtain

$$b_{k'}=-\oint \beta^P_{l'k'}\,dx_{l'}=-\int_s \partial_{[j'}\beta^P_{l']k'}\,dF_{j'l'} \tag{9.5}$$

or alternately as

$$b_{k'}=-\oint \beta^P_{l'k'}\,dx_{l'}=-\int_s \varepsilon_{i'j'l'}\partial_{j'}\,\beta^P_{l'k'}\,dF_{i'} \tag{9.6}$$

where $\varepsilon_{i'j'l'}$ is the permutation symbol in Cartesian coordinates. In the case of Eq. 9.5, we obtain

$$b_2=-\int_s \partial_1\,\beta^P_{22}\,dF_{12}=\int_s V\delta(x_1)\,dF_{12}=4V \tag{9.7}$$

which is the same result as that given by Eq. 9.4. Eq. 9.6 also leads to this same result. Similar to Eq. 4.61 for the nonlinear expression, we can write for the linearized form

$$b_{k'}=-\int_s S_{j'l'k'}\,dF_{j'l'} \tag{9.8}$$

where the linearized form of the torsion tensor is given by

$$S_{j'l'k'}=\partial_{[j'}\beta^P_{l']k'} \tag{9.9}$$

which yields

$$S_{122}=\left\{-\left(\tfrac{1}{2}\right)V\delta(x_1)\right\}_1 \tag{9.10}$$

and can be compared to the nonlinear form given by Eq. 4.78. Along the same lines as that done in Eq. 4.115, Eq. 9.8 can be rewritten as

$$db_{k'}=-S_{j'l'k'}\,dF_{j'l'}=-S_{j'l'k'}\varepsilon_{i'j'l'}\,dF_{i'}=\alpha_{i'k'}\,dF_{i'} \tag{9.11}$$

where $\alpha_{i'k'}$ is the dislocation density tensor, analogous to that given by Eq. 4.116, but expressed in linear form, that is,

$$\alpha_{i'k'} = -\varepsilon_{i'j'l'} S_{j'l'k'} \tag{9.12}$$

It also follows from Eqs. 9.6 and 9.11 that

$$b_{k'} = \int_s \alpha_{i'k'} \, dF_{i'} \tag{9.13}$$

so that the Burgers vector can be obtained directly from the dislocation density tensor. For the linearized version of the (k') state shown in Figure 4.2, Eq. 9.12 yields

$$\alpha_{32} = -2S_{122} = \{V\delta(x_1)\}_1 \tag{9.14}$$

which when integrated with respect to x_1, yields the value of V, and simply represents the number of extra half planes per plane of the original undistorted crystal along the x_2 direction. In other words, α_{32} is a dislocation density per unit length. The simplicity afforded by the linearized theory is thus rather obvious, and has been treated in somewhat greater detail by Kröner (1964). Equation 9.12 could also be written in alternate form as

$$\boldsymbol{\alpha} = -\mathrm{cur}\,l\,\boldsymbol{\beta}^P \tag{9.14}$$

from which it follows that

$$\mathrm{div}\,\mathrm{cur}\,l\,\boldsymbol{\beta}^P = 0 \tag{9.15}$$

This last equation simply means that dislocations cannot end within the crystal.

The $(k^{1'})$ state shown in Figure 4.3 can also be treated in the limit of the linear approximation by letting V in Eq. 9.3 be a negative number whose absolute value is much less than unity. Otherwise, the analysis follows the same as that for the (k') state.

For the linear approximation to the $(k^{2'})$ state shown in Figure 4.5, Eq. 4.18b yields for the distortion tensor $\beta^P_{k^{2'}l^{2'}}$

$$\beta^P_{21} = \{\theta H(-x_1)\}_1 \tag{9.16}$$

that when used in conjunction with the following relation

$$b_{k^{2'}} = -\oint \beta^P_{k^{2'}l^{2'}} \, dx_{l^{2'}} \tag{9.17}$$

yields

$$b_1 = -\int_3^4 \theta\, dx_2 = 4\theta \tag{9.18}$$

Alternatively, we could also write

$$b_{k^{2'}} = -\int_s S_{j^{2'}l^{2'}k^{2'}}\, dF_{j^{2'}l^{2'}} \tag{9.19}$$

and since

$$S_{121} = \left\{ -\left(\tfrac{1}{2}\right)\theta\,\delta(x_1)\right\}_1 \tag{9.20}$$

Eq. 9.19 yields the same result as Eq. 9.18. The dislocation density tensor for the $(k^{2'})$ state is simply

$$\alpha_{31} = -2S_{121} = \theta \tag{9.21}$$

For the single dislocation associated with the (k) state of Figure 4.1*b*, a relation of the type given by Eq. 9.1 along with Eq. 4.2b yields for β_{kl}^{P}

$$\beta_{22}^{P} = H(-x_1)\,\delta(x_2) \tag{9.22}$$

which when substituted into the following equation

$$b_k = -\oint \beta_{kl}^{P}\, dx_l \tag{9.23}$$

yields

$$b_2 = -\int_3^4 \delta(x_2)\, dx_2 = 1 \tag{9.24}$$

We could also write

$$b_k = -\int_s S_{jlk}\, dF_{jl} \tag{9.25}$$

and since

$$S_{122} = \left\{ -\left(\tfrac{1}{2}\right)\delta(x_1)\,\delta(x_1)\right\}_1 \tag{9.26}$$

Eq. 9.25 gives the same result as Eq. 9.24. Finally, for the (k) state

$$\alpha_{32} = -2S_{122} = 1 \tag{9.27}$$

where the double integration over x_1 and x_2 simply means that the dislocation density has no units, but is rather a pure number, that is, one dislocation. A similar analysis follows for the (k^1) state of Figure 4.3*b* and the (k^2) state of Figure 4.4*b*, except that for $\beta^P_{k^1l^1}$, we must write

$$\beta^P_{32} = -H(-x_1)\delta(x_2) \tag{9.28a}$$

while for $\beta^P_{k^2l^2}$

$$\beta^P_{11} = H(+x_2)\delta(x_1) \tag{9.28b}$$

The stress-free plastic or anholonomic states considered in Section 5.1 can also be easily extended to the realm of the linear approximation. In particular, the linearized form of Eq. 5.44 becomes

$$b_a = -\int_S \Gamma_{cba}\, dF_{cb} = -\int_S (S_{cba} - \Omega_{cba})\, dF_{cb} \tag{9.29}$$

where similar to Eq. 9.9, we can write

$$S_{cba} = \partial_{[c}\beta^P_{b]a} \tag{9.30}$$

while

$$\Omega_{cba} = S_{cba} \tag{9.31}$$

It is apparent that the only nonvanishing component of S_{cba} is S_{122} and is identical to that given by Eq. 9.10. The analysis for b_a thus follows along the same lines as that for the (k') state. In addition, however, we can obtain a newly created free surface density similar to that given by Eq. 5.77 as

$$\beta_{da} = \varepsilon_{dcb}\Omega_{cba} \tag{9.32}$$

In considering the remaining plastic distortions corresponding to the linearized counterparts of Section 5.1, we find that analogous to the elastic distortions given by Eq. 8.25a, we can decompose a simple plastic shear into a pure plastic shear plus a pure plastic rotation as follows:

$$\beta^P_{a^{2'}b^{2'}} = \beta^{Pe}_{a^{2'}b^{2'}} + \beta^{P\omega}_{a^{2'}b^{2'}} \tag{9.33a}$$

where the three terms are identical to those given by Eq. 8.25b. Also similar to Eq. 8.25c, we can write

$$\beta^P_{a^{6'}b^{6'}} = 2\beta^{Pe}_{a^{2'}b^{2'}} \tag{9.33b}$$

while

$$\beta^P_{a^{4'}b^{4'}} = \beta^P_{a^{3'}b^{3'}} = \begin{bmatrix} 0 & -\theta & 0 \\ \theta & 0 & 0 \\ 0 & 0 & 0 \end{bmatrix} \tag{9.33c}$$

which is the plastic counterpart of the elastic distortions given by Eq. 8.26. The last two equations must be used with their Heaviside functions $H(-x_a)$. It is a straightforward matter to write a set of equations for these states similar to that given by Eqs. 9.29 and 9.32.

In linearized form, the equation of parallel displacement given by Eq. 5.32 becomes

$$dC_k = -\Gamma_{mlk} C_l \, dx_m \tag{9.34}$$

Also, from Eq. 5.23, we may write

$$S_{mlk} = \tfrac{1}{2}[\Gamma_{mlk} - \Gamma_{lmk}] \tag{9.35}$$

or in view of Eq. 9.9

$$S_{mlk} = \partial_{[m}\beta^P_{l]k} \tag{9.36}$$

so that

$$\Gamma_{mlk} = \partial_m \beta^P_{lk} \tag{9.37}$$

Considering the linearized form of the (k') state shown in Figure 4.2, Eq. 9.34 becomes

$$dC_2 = -\Gamma_{122} C_2 \, dx_1 \tag{9.38}$$

and from Eqs. 9.35 and 9.10

$$\Gamma_{122} = 2S_{122} = \{-V\delta(x_1)\}_1 \tag{9.39}$$

so that

$$dC_2 = VC_2 \delta(x_1) \, dx_1 \tag{9.40}$$

Thus, in terms of Figure 5.7*a*, when a vector $C_2 \equiv dx_{\underset{2}{k}}$ is displaced parallel to the left, it changes its length by VC_2 and corresponds to the dotted arrow. Similarly, in the linearized case of the $(k^{2'})$ state of Figure 4.5 we may write

$$dC_1 = -\Gamma_{121} C_2 dx_1 \tag{9.41}$$

Utilizing Eqs. 9.35 and 9.20 we find

$$\Gamma_{121} = 2S_{121} = \{-\theta\delta(x_1)\}_1 \tag{9.42}$$

so that

$$dC_1 = \theta C_2 \delta(x_1)\, dx_1 \tag{9.43}$$

Thus, in terms of Figure 5.7*b*, when a vector $C_2 \equiv dx_{\underset{2}{k^2}}$ is displaced parallel to the left, it changes its length by θC_2 corresponding to the dotted arrow.

It is now a simple matter to apply the linear approximation to the surface dislocation arrays discussed in Section 5.2. In particular, the coordinates remain Cartesian so that the Burgers vectors of the various arrays remain parallel or perpendicular in these coordinates, as is in fact schematically illustrated in Figures 5.18, 5.22*b*, 5.25*b*, 5.28*b*, and 5.31*b*.

We next proceed to the linearization of the grain boundary and two-phase interface problem treated in Section 6. In particular, for the linearized case of the (k^{I1}) state shown in Figure 6.1*a*, Eq. 6.1 can be written as

$$\beta^P_{12} = e^P_{12} = \{H(-x_1)\theta\}_1 + \{H(+x_1)\theta\}_2 \tag{9.44}$$

Also, since it is no longer necessary to distinguish between the coordinates within each grain since both belong to the same Cartesian system, we can alternately write Eq. 9.44 as

$$\beta^P_{12} = \{2H(-x_1)\theta\}_1 = \{-2H(+x_1)\theta\}_2 \tag{9.45}$$

which can be substituted into the following equation to obtain the grain boundary Burgers vector

$$b_{k^{II'}} = -\oint \beta^P_{k^{II'} l^{II'}} dx_{l^{II'}} \tag{9.46}$$

Equation 9.44 can also be used to obtain the torsion tensor

$$S_{121}=\partial_{[1}\beta^{P}_{2]1}=\{-(\tfrac{1}{2})\theta\,\delta(x_1)\}_1+\{-(\tfrac{1}{2})\theta\,\delta(x_1)\}_2\equiv\{-\theta\,\delta(x_1)\} \tag{9.47}$$

Note that since it is no longer necessary to distinguish between coordinates in the two grains, the two bracketed terms can be added. This result is also the same as what would be obtained if Eq. 9.45 were used in conjunction with Eq. 9.47. We can now write

$$b_1=-\int_s S_{121}\,dF_{12}=2\theta\,\Delta x_2=8\theta \tag{9.48}$$

where Δx_2 has been chosen to be the same length as that for the nonlinear case that led to Eq. 6.3b, that is, 4. Since Eq. 9.46 yields only the component of the vector **b**, we can write, similar to Eq. 6.6

$$\mathbf{b}=8\theta\mathbf{e}_1 \tag{9.49}$$

where $\mathbf{e}_1$ is the base vector corresponding to Cartesian coordinates. The grain boundary dislocation density can now be written as

$$\alpha_{31}=-2S_{121}=2\theta \tag{9.50}$$

This value simply corresponds to the total b_1 component of the Burgers vectors within the boundary per unit length of boundary. Again, the interpretation is simplified compared to that of the nonlinear case given by Eq. 6.9 because the grain boundary, as well as the two adjacent grains, now have the same reference frame as that of the original undistorted crystal.

Since in view of Eq. 9.33a, a pure shear consists of both a simple shear plus a rigid rotation, the distortion given by Eq. 9.44 could have been alternately written as

$$\beta^{P}_{12}=\beta^{Pe}_{12}+\beta^{P\omega}_{12}=\left\{\left[\frac{\theta}{2}H(-x_1)\right]+\left[\frac{\theta}{2}H(-x_1)\right]\right\}_1 +\left\{\left[-\frac{\theta}{2}H(+x_1)\right]+\left[-\frac{\theta}{2}H(+x_1)\right]\right\}_2 \tag{9.51a}$$

and

$$\beta^{P}_{21}=\beta^{Pe}_{21}+\beta^{P\omega}_{21}=\left\{\left[\frac{\theta}{2}H(-x_1)\right]+\left[-\frac{\theta}{2}H(-x_1)\right]\right\}_1 +\left\{\left[-\frac{\theta}{2}H(+x_1)\right]-\left[-\frac{\theta}{2}H(+x_1)\right]\right\}_2 \tag{9.51b}$$

The linearized form of the anholonomic (a^{I1}) state shown in Figure 6.1b is also rather easily obtained from Eq. 6.14 where we write

$$\Omega_{121} = S_{121} \tag{9.52}$$

It is also clear from Eqs. 6.21 and 6.24 that the linearized forms of states (k^{I3}) and (k^{I4}), respectively, shown in Figures 6.1d and 6.3 become identical, whereas $\beta_{12}^P = -\beta_{21}^P$ for these states. In addition, β_{12}^P for the linearized (k^{I3}), (k^{I4}), and (k^{I1}) all become identical, that is, namely that given by Eq. 9.44, so that the grain boundary configurations shown in Figures 6.1a, 6.1d, and 6.3 become the same. On the other hand, the total distortion associated with these states is not the same, since $\beta_{21}^P = 0$ for the ($k^{I1'}$) state. It has sometimes been argued that only the distortions given by $\beta_{12}^{P\omega}$ in Eq. 6.51a are needed to describe the dislocation content of a grain boundary. This is obviously incorrect since both simple plastic shear and plastic rotation each contribute their own dislocation contents.

Turning now to the linearized form of the (k^{I5}) state shown in Figure 6.4, it is clear from Eq. 6.35 that the distortions become infinite so that the linear approximation breaks down for this case. What this means physically is that for a grain boundary with the atom configuration shown in Figure 6.4 the dislocation configuration shown in Figure 6.1a will perhaps be the more acceptable one. From an energy point of view, this latter configuration would be more favorable since it would be of much shorter length. This same reasoning also holds for the (k^{I6}) state of Figure 6.5. On the other hand, when the volume expansion is removed from this state, we obtain the rigid plastic rotation shown in Figure 6.6 and given by Eq. 6.47. In linearized form, Eq. 6.47 becomes

$$\underset{1}{A}{}_K^{k^{I7'}} = \begin{bmatrix} -\theta & 1 & 0 \\ -1 & -\theta & 0 \\ 0 & 0 & 1 \end{bmatrix} \tag{9.53}$$

This is a distortion which simply interchanges the roles of the x_1 and x_2 axes and thus again does not properly represent the correct distortion, again because of the large rotations involved.

For the linearized counterpart of the (k^{I9}) state of Figure 6.7b, it follows from Eq. 6.60, that

$$\beta_{12}^P = \{H(-x_1)\theta\}_1 \tag{9.54a}$$

while

$$\beta_{22}^P = \left\{ \frac{1-\cos\theta}{\cos\theta} H(+x_1) \right\}_2 \cong \{(1-\cos\theta)H(+x_1)\}_2 \tag{9.54b}$$

The same results also obtain for the linearized form of the asymmetric tilt boundary formed by rigid rotation as shown in Figure 6.7*c*. The above distortions yield nonvanishing components of the torsion tensor given by

$$S_{121}=\alpha_{[1}\beta^{P}_{2]1}=\{-\theta\delta(x_1)\} \tag{9.55a}$$

and

$$S_{122}=\partial_{[1}\beta_{2]2}=\{(1-\cos\theta)\delta(x_1)\} \tag{9.55b}$$

which when substituted into the following equations:

$$b_{k^{I9'}}=-\int_s S_{j^{I9'}l^{I9'}k^{I9'}}\,dF_{j^{I9'}l^{I9'}} \tag{9.56a}$$

and

$$\alpha_{i^{I9'}k^{I9'}}=-\varepsilon_{i^{I9'}j^{I9'}l^{I9'}}S_{j^{I9'}l^{I9'}k^{I9'}} \tag{9.56b}$$

yield the Burgers vector and dislocation density associated with the linearized approximation of the boundary shown in Figures 6.7*b* and 6.7*c*.

Similar to the case shown for the (k^{1c}) state illustrated in Figure 4.6, it is possible to represent a symmetric type low angle grain boundary such as that shown in Figure 9.1*a* in terms of another coordinate system such as illustrated in Figure 9.1*b*. However, unlike the (k^{1c}) state, we must now restrict ourselves to transformations that are Cartesian in nature, that is, orthogonal transformations. Such would be the case for the rigid rotation ϕ given by the inverse of Eq. 2.27. We could thus write

$$\bar{b}_{k^{I1'}}=C_{k^{I1'}l^{I1'}}b_{l^{I1'}} \tag{9.57}$$

where $\bar{b}_{k^{I1'}}$ is the Burgers vector associated with the $(k^{I1c'})$ state of Figure 9.1*b*. Using the inverse of Eq. 2.27, we find

$$\bar{b}_1=\cos\phi\, b_1 \tag{9.58a}$$

while

$$\bar{b}_2=-\sin\phi\, b_1 \tag{9.58b}$$

where θ in Figure 9.1*b* was chosen as 45°. Again, it should be emphasized that Figure 9.1*b* does not necessarily mean that the dislocations in Figure 9.1*a* dissociate into the array shown in Figure 9.1*b*, but rather that they are

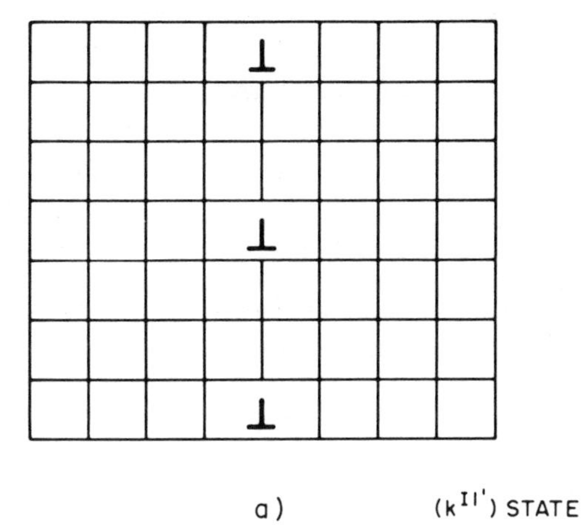

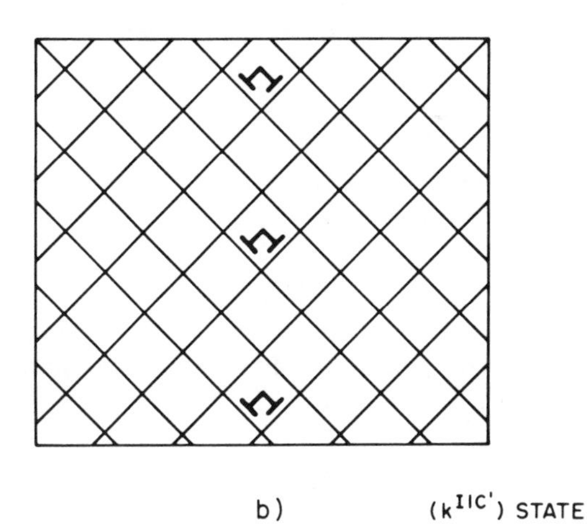

Figure 9.1 Representation of low-angle grain boundary shown in *a*) in terms of another coordinate system *b*).

merely different coordinate representations of the same array. A dissociation could of course occur if a corresponding energy decrease where to occur, but this would be another matter. Other tensor quantities such as the torsion tensor and dislocation density could also be transformed similar to Eq. 9.57 with equations of the form

$$\bar{S}_{ijk} = C_{il} C_{jm} C_{kn} S_{lmn} \tag{9.59}$$

that in view of Eq. 9.47 yields

$$\bar{S}_{121}=\{-\theta\cos\phi\,\delta(x_1)\} \tag{9.60a}$$

and

$$\bar{S}_{122}=\{\theta\sin\phi\,\delta(x_1)\} \tag{9.60b}$$

that when integrated over the plane 1–2 yields the same result as that given by Eq. 9.58. Similarly, we can also write

$$\bar{\alpha}_{ij}=C_{ik}C_{jl}\alpha_{kl} \tag{9.61}$$

that in view of Eq. 9.50 yields

$$\bar{\alpha}_{31}=2\theta\cos\phi \tag{9.62a}$$

and

$$\bar{\alpha}_{32}=-2\theta\sin\phi \tag{9.63a}$$

Turning now to the linearized case of the (k^{I10}) state shown in Figure 6.9*c*, we can write for the distortion using Eq. 6.74b

$$\beta^P_{22}=\{(V-1)H(+x_1)\}_2 \tag{9.64}$$

where $(V-1)$ is much less than unity. Eq. 9.64 is thus similar to Eq. 9.3. The torsion tensor can be written as

$$S_{122}=\partial_{[1}\beta^P_{2]2}=\{(\tfrac{1}{2})(V-1)\delta(x_1)\}_2 \tag{9.65}$$

from which both the Burgers vector and dislocation density of the two-phase interface can be determined.

For the linearized form of the (κ^I) state of Figure 6.8*a*, Eq. 6.87 can be rewritten as

$$Q_{mlk}=-2\partial_m e_{lk} \tag{9.66}$$

and since

$$e_{lk}=\tfrac{1}{2}(g_{lk}-\delta_{lk}) \tag{9.67}$$

where

$$g_{11} = g_{33} = 1 \tag{9.68a}$$

and

$$g_{22} = \{(1+V)^2 H(-x_1)\}_1 + \{H(+x_1)\}_2 \tag{9.68b}$$

we have

$$e_{22} = \tfrac{1}{2}[\{(1+2V)H(-x_1)\}_1 + \{H(+x_1)\}_2 - 1] \tag{9.69}$$

that when substituted into Eq. 9.66 yields

$$Q_{122} = -2\partial_1 e_{22} = 2V\delta(x_1) \tag{9.70}$$

that can be substituted into the linearized form of Eq. 6.95 to obtain

$$b_2 = \frac{1}{2}\int_S Q_{122}\, dF_{12} = V\Delta x_2 = 4V \tag{9.71}$$

In addition, the dislocation density is easily found from the linearized form of Eq. 6.98, that is,

$$\alpha_{nk} = \tfrac{1}{2} e_{nml} Q_{mlk} \tag{9.72}$$

that in view of Eq. 9.70 yields after integration with respect to x_1

$$\alpha_{32} = Q_{122} = V \tag{9.73}$$

REVIEW

The various tensor quantities presented in their most general nonlinear forms in Chapters 4 and 5 have been linearized in this chapter. This linearization leads to great simplification of these quantities by allowing one to maintain Cartesian coordinates. For example, Eq. 9.2 gives the linearized form of the line integral that yields the Burgers vector and corresponds to the nonlinear form given by Eq. 4.37 in Chapter 4. Similarly, the linearized expressions for the torsion tensor, anholomomic object, dislocation density tensor, parallel displacement of a vector, Q

tensor, and quasidislocation density are given by Eqs. 9.9, 9.31, 9.12, 9.34, 9.66, and 9.72, respectively, and correspond to the more general nonlinear forms given by Eqs. 4.62, 5.38, 4.117, 5.32, 6.87, and 6.98, respectively. Equation 9.8 shows how the linearized torsion can be integrated with respect to a given area to yield the Burgers vector and corresponds to the more general form given by Eq. 4.61.

The various dislocation arrays considered in their more general forms in Chapters 4 and 5 have all been reanalyzed in their simpler linear forms in the present chapter. Likewise, the corresponding tensor quantities associated with the more generalized internal boundaries in Chapter 6 have been reduced to their linear forms. As is shown in Chapter 13, it is in its linear form that dislocation theory has its greatest and widest application to a vast realm of practical problems. It should also be pointed out here that all of the developments in Chapters 8 and 9 have concerned themselves with geometrical considerations, of which strain is a particular outcome of this analysis. The constituitive relations between stress and strain in the linear approximation is the subject of Chapter 11.

CHAPTER 10

Screw-Type Dislocation Arrays

Up until now, all of the dislocation arrays considered have been of the edge-type, that is, those with Burgers vectors normal to the dislocation line. In this chapter we now wish to turn our attention to that class of dislocations whose Burgers vector lies parallel to the dislocation, that is, screw-type arrays. Perhaps the most important configurations that are characterized by such arrays are twist-type grain boundaries. In addition, we are now in a position to completely describe any state of either plastic or elastic strain in a three-dimensional body in terms of a combination of screw and edge-type arrays.

Thus far, all of the dislocation configurations examined, that formed interfaces, involved edge type dislocations. It is instructive at this point to consider a whole new class of interfaces which involve screw type dislocations, that is, dislocations in which the Burgers vector is parallel to the dislocation line. This is best done by considering the class of distortions such as those discussed in Section 5 that generated a boundary between elastically and plastically strained regions. In particular, Figure 10.1*a* consists of an interface that lies in the plane of the drawing and that separates regions of pure elastic and plastic shear similar to that shown in Figure 5.24*a*. It is seen, however, that this time the interface consists of a parallel array of screw type dislocations that are drawn dotted. Also, for clarity, it is seen that the screw dislocations terminate on edge dislocations similar to those shown in Figure 5.24*a*. For convenience, the elastically deformed region #1 may be viewed as lying below the plane of the drawing, while the plastically distorted region #2 lies above this plane. It also follows from Figure 5.23 that Figure 10.1*a* represents the dislocation

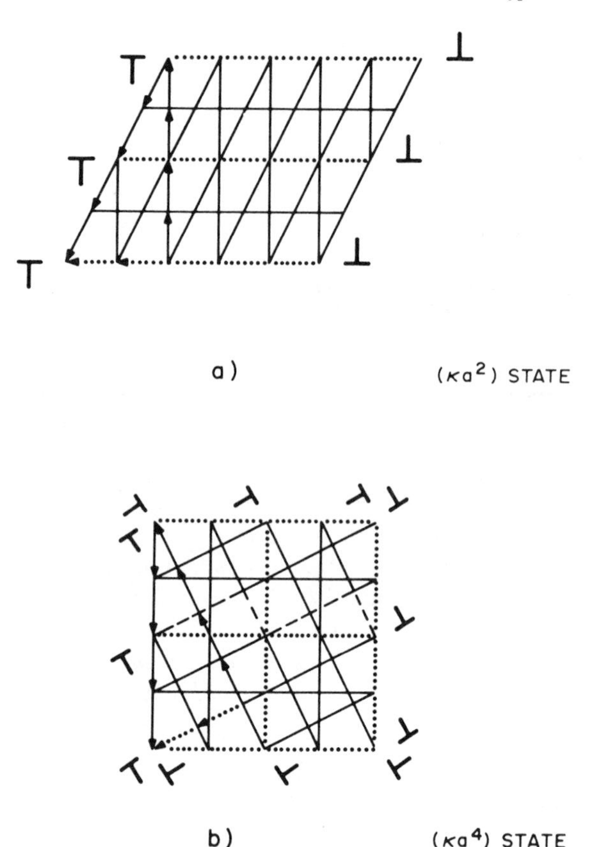

Figure 10.1 *a*) Elastically and plastically sheared regions analogous to those shown in Figure 5.24*a* that form an interface that now lies in the plane of the drawing. *b*) Doubly sheared regions similar to those shown in Figure 5.27*a* that form an interface that now lies in the plane of the drawing.

configuration on the face of a rectangular hole which is finite in the z direction, that is, normal to the plane of a drawing. In this manner, the dislocation configuration about all six faces of a rectangular body could be determined that would be more general than the two-dimensional cases depicted in Figures 5.25*a* and *b*.

If now the doubly sheared plastic and elastic regions of Figure 5.27*a* were joined along the plane of the drawing, the configuration shown in Figure 10.1*b* would obtain. The dotted lines represent an orthogonal cross grid of dislocations that lie in the plane of the drawing and that terminate on pure edge type dislocations. These dotted lines can, in fact, be visualized as being comprised of orthogonal segments of dislocations that lie along the axes of the plastically deformed crystals. One such set of segments is shown dashed in Figure 10.1*b* and corresponds to the horizontal dotted line passing through the center of the figure. A clearer understanding of this construction may be obtained by referring to Figure 10.2 which corresponds to the face of the plastically deformed region of Figure 10.1*b*. The distortion associated with the double shear may be viewed in terms of the creation of numerous edge dislocation dipoles within the crystal. These dipoles terminate on vacancies or holes, designated by shaded areas, that in fact give rise to the volume increase associated with this particular distortion. These holes may be filled with matter with no loss in generality. It is also evident that each dipole has associated with it a screw dislocation segment that lies in the plane of the drawing. Furthermore, certain sets of dislocation dipoles may be connected such as shown by the stepped dotted line in Figure 10.2. This stepped segment corresponds to the dashed segment in Figure 10.1*b*. In summary, the dashed line in this figure may be viewed as the dislocation representation in terms of the original crystal lattice, that is, crystal lattice dislocations, whereas the dotted line representation may be viewed as the dislocation representation in terms of a coincidence site lattice, that is, boundary dislocations, as discussed in Section 7. Unlike the case discussed with respect to Figure 10.1*a*, the dislocations in Figure 10.1*b* are not of pure screw type but also contain edge components. These edge components may be viewed as that

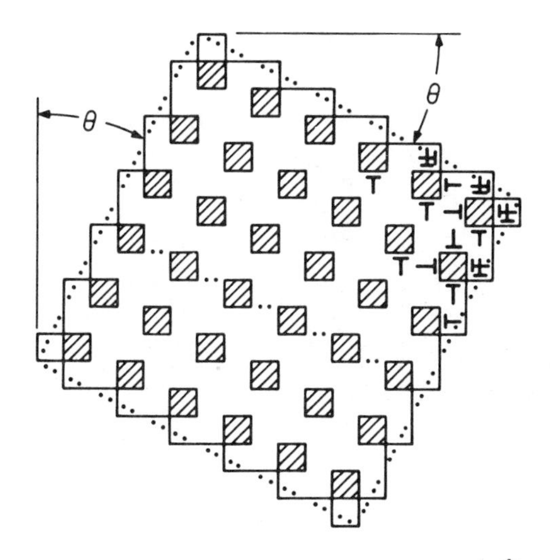

Figure 10.2 More detailed description of the doubly sheared plastic state shown in Figure 3.3*a*.

contribution from the cross grid of dislocations that is associated with the volume change accompanying the distortions in Figure 10.1*b*.

When the pair of crystals in Figure 5.29*a* showing pure rigid and plastic rotations are superimposed upon one another, the interface shown in Figure 10.3*a* obtains. As was the case in Figure 10.1*b*, an orthogonal grid of dislocations is observed that have both screw and edge components. However, the cross grid, or conversely the edge dislocations upon which they terminate, are more closely spaced than those shown in Figure 10.1*b*.

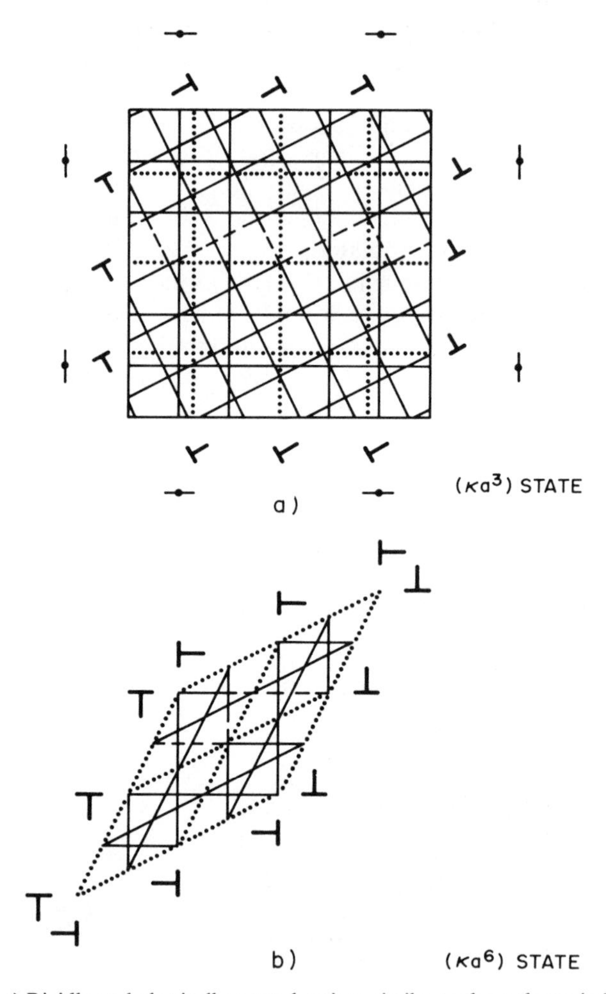

Figure 10.3 *a*) Rigidly and plastically rotated regions similar to those shown in Figure 5.29*a*, but that now form an interface that lies in the plane of the drawing. *b*) Simply sheared regions similar to those shown in Figures 2.4 and 3.4 that form an interface that lies in the plane of the drawing.

This is related to the fact that the bottommost region in Figure 10.3*a* may be viewed as comprised of a cross grid of prismatic type dislocations. This is apparent from inspection of the leftmost region of Figure 5.29*a*. These prismatic dislocations may be visualized as arising from the elastic relaxation, that is, contraction of the bottom region of Figure 10.1*b* and the corresponding plastic contraction of the topmost half of this body. In other words, extra matter is removed from the topmost region of Figure 10.1*b* that is equivalent to the addition of prismatic dislocations to the lower half of this body.

Finally, for completeness, Figure 10.3*b* shows an interface formed by the superposition of elastic and plastic simply sheared bodies such as depicted in Figures 2.4 and 3.4, respectively. A cross grid of dotted dislocations is again generated, as was the case in Figure 10.1*b*. This time, however, the grid is not orthogonal, but once again the dislocations possess both screw and edge type character. In addition, the Burgers vectors of the nonvertical dotted arrays have opposite sign from those in Figure 10.1*b*. Again, the dotted arrays, that may be termed boundary dislocations, are seen to be comprised of stepped dislocation segments that lie along the original crystal lattice directions. One such stepped array is shown dashed in Figure 10.3*b* and corresponds to the nonhorizontal dotted line passing through the center of the body. The nature of these dislocations is seen more clearly by reference to Figure 10.4 that shows in more detail the nature of the plastically deformed portion of Figure 10.3*b*. Once again, as in Figure 10.2, the distortion may be described in terms of an array of dislocation dipoles. This time, however, the dipoles terminate on interstitial or extra matter that is denoted by the shaded areas. Each dipole also bounds a segment of screw dislocation and the dotted stepped segment in Figure 10.4 shows how the dipoles are connected to generate a single dislocation line as in Figure 10.3*b*.

Returning now to Figure 10.1*a*, it is apparent that a Burgers circuit can be constructed in the manner prescribed by the arrows. It is similar to that shown in Figure 5.24*a* except that, in general, part of its path must be taken along the z direction. This is shown more clearly in Figure 10.5 which represents a three-dimensional portrayal of Figure 10.1*a*. In any event, both circuits lead to the same closure failure. Again we denote this state as (κa^2) and define a distortion tensor $A_K^{\kappa a^2}$ similar to Eq. 5.101, but modified to read

$$A_1^1 = A_2^2 = A_3^3 = 1 \tag{10.1a}$$

and

$$A_2^1 = \{0\}_1 + \{H(+x^3)\tan\theta\}_2 \tag{10.1b}$$

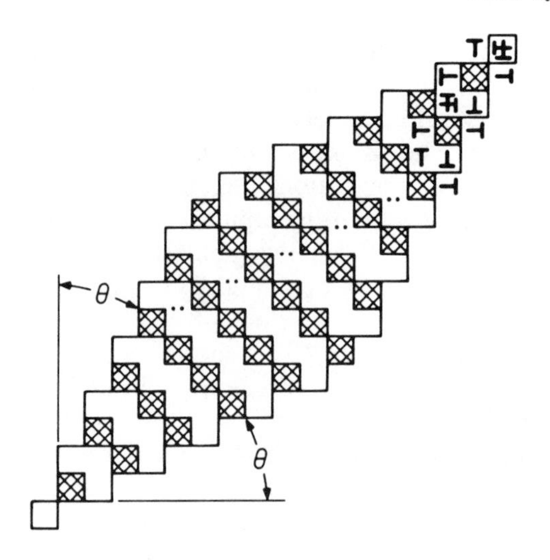

Figure 10.4 More detailed description of the simply sheared plastic state shown in Figure 3.4.

where $+x^3$ refers to the normal direction above the plane of the drawing. Equations 5.102 and 5.103 can be used to determine the Burgers vector of Figure 10.1a. In the latter case we find for the torsion tensor

$$S_{23}^{\cdot\cdot 1} = -\tfrac{1}{2}\bar{A}_2^2\bar{A}_3^3\partial_3 A_2^1 = \{0\}_1 + \left\{-\left(\tfrac{1}{2}\right)\delta(x^3)\tan\theta\right\}_2 \tag{10.2}$$

that when substituted into Eq. 5.103 yields

$$b^1 = -4\tan\theta = -2 \tag{10.3}$$

The negative sign in the above equation arises from the fact that dF^{23} in Eq. 5.103 is negative since the outward pointing normal to the leftmost face of Figure 10.5 is in the negative direction. Equation 10.2 can also be used to obtain the following dislocation density

$$\alpha^{11} = -2S_{23}^{\cdot\cdot 1} = \delta(x^3)\tan\theta \tag{10.4}$$

This simply corresponds to the amount of displacement in the x^1 direction per unit length along x^2 caused by the screw dislocations located at the interface shown dotted in Figure 10.5. Note that for screw dislocations the dislocation density is always given by the components α^{kk}.

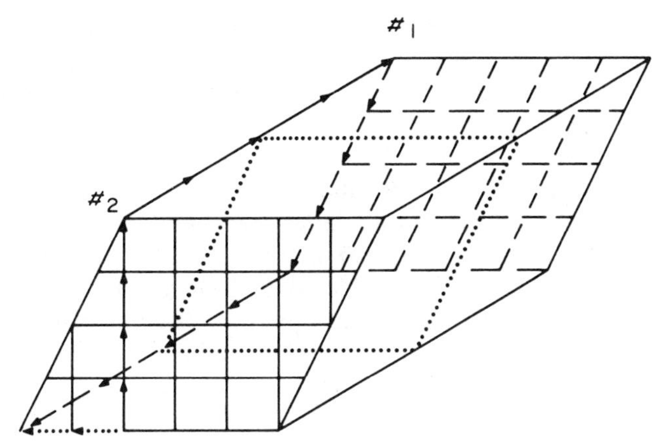

Figure 10.5 Three-dimensional view of Figure 10.1*a* to demonstrate construction of the Burgers circuit.

For the (κa^4) state shown in Figure 10.1*b* the nonvanishing components of the distortion tensor become

$$A_1^1 = A_2^2 = A_3^3 = 1 \qquad (10.5a)$$

and

$$A_2^1 = -A_1^2 = \{0\}_1 + \{H(+x^3)\tan\theta\}_2 \qquad (10.5b)$$

These results are similar to those given by Eq. 10.1 except that they now yield a second nonvanishing component of the torsion tensor given by

$$S_{13}^{\cdot\cdot 2} = -\tfrac{1}{2}\bar{A}_1^1\bar{A}_3^3\partial_3 A_1^2 = \{0\}_1 + \left\{\left(\tfrac{1}{2}\right)\delta(x^3)\tan\theta\right\}_2 \qquad (10.6)$$

that can be substituted into Eq. 5.103 to yield

$$b^2 = 4\tan\theta = -2 \qquad (10.7)$$

We can also obtain for the corresponding component of the dislocation density tensor

$$\alpha^{22} = 2S_{13}^{\cdot\cdot 2} = \delta(x^3)\tan\theta \qquad (10.8)$$

The component $S_{23}^{\cdot\cdot 1}$ yields the dotted arrows associated with the Burgers circuit located at the left of Figure 10.1*b*. In particular, it is simply the displacement in the x^1 direction of all those dislocations that lie along a given length x^2. When x^2 is chosen as unit length, the displacement along x^2 yields the dislocation density α^{11} where the first 1 in the superscript α^{11} corresponds to the direction of the dislocation line. Strictly speaking, α^{11} applies to a continuous distribution of dislocations in which the discrete configuration shown in Figure 10.2 may be visualized as subdivided into a continuous distribution. However, the present discrete presentation leads to no conceptual difficulties. In a similar manner, $S_{13}^{\cdot\cdot 2}$ asnd α^{22} given by Eqs. 10.6 and 10.8 can be used to find the closure failure normal to that given in Figure 10.1*b*, but that for reasons of clarity are not shown. The quantities $b^{\kappa a^4}$ and $\alpha^{\kappa d^4 \kappa a^4}$ can also be expressed in terms of the boundary coordinates shown dotted in Figure 10.1*b* by writing

$$b^{I\kappa a^4} = C_{\kappa a^4}^{I\kappa a^4} b^{\kappa a^4} \tag{10.9}$$

where $C_{\kappa a^4}^{I\kappa a^4}$ is simply a coordinate rotation that makes the inclined orthogonal grid in Figure 10.1*b* correspond to the orientation of the dotted one. A similar result can also be written for $\alpha^{I\kappa d^4 I\kappa a^4}$.

Turning next to the (κa^3) state of Figure 10.3*a*, we see that it is also equivalent to the asymmetric tilt boundary of Figure 6.7*c*. The distortions given by Eq. 10.5 must therefore be modified to read

$$A_3^3 = 1 \tag{10.10a}$$

$$A_1^1 = A_2^2 = \{0\}_1 + \{H(+x^3)\cos\theta\}_2 \tag{10.10b}$$

$$A_2^1 = A_1^2 = \{0\}_1 + \{H(+x^3)\sin\theta\}_2 \tag{10.10c}$$

These in turn can be used to obtain nonvanishing components of the torsion tensor given by $S_{13}^{\cdot\cdot 1}$, $S_{23}^{\cdot\cdot 2}$, $S_{23}^{\cdot\cdot 1}$, and $S_{13}^{\cdot\cdot 2}$. The first two components yield the Burgers vectors associated with the outermost prismatic dislocation arrays shown in Figure 10.3*a*.

Finally, for the (κa^6) shown in Figure 10.3*b*, the distortions given by Eq. 10.5 must be modified as follows:

$$A_1^1 = A_2^2 = A_3^3 = 1 \tag{10.11a}$$

$$A_2^1 = -A_1^2 = \{0\}_1 + \{H(+x^3)\tan\theta\}_2 \tag{10.11b}$$

from which the torsion and dislocation density tensors can easily be found.

We are now in a position to construct the twist boundary counterparts of the tilt boundaries treated in Section 6. In particular, Figure 10.6*a* shows a twist boundary formed by the superposition of the pair of grains shown in Figure 6.1*a*. In order to make the grains coherent at the boundary, analogous to that which was described in connection with

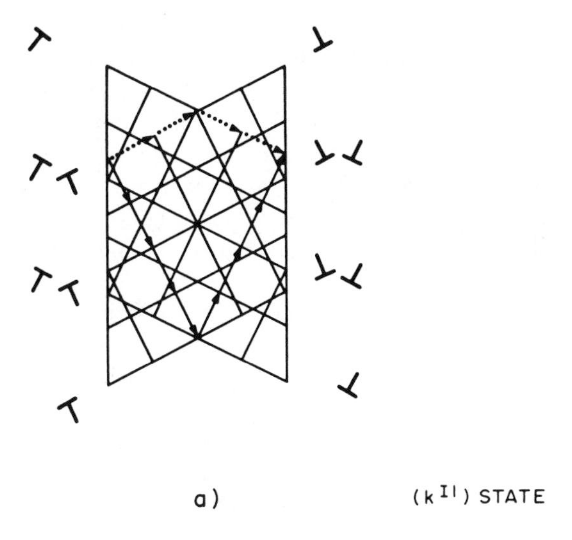

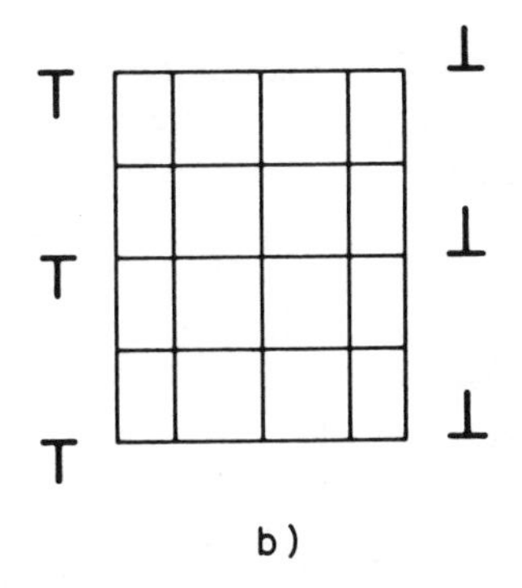

Figure 10.6 Twist boundary counterpart of symmetric tilt boundary shown in Figure 6.1*a*. *a*) Configuration of two grains away from boundary. *b*) Configuration right at boundary.

Figure 6.1, they must be distorted to a common lattice such as shown in Figure 10.6*b*. The resulting configuration may be described as an array of horizontal parallel screw dislocations that terminate on the edge dislocations shown in Figure 10.6*b*. In effect, the Burgers vectors associated with these dislocations are the vector sums of the crystal lattice dislocations shown dotted in Figure 10.6*a*.

If, on the other hand, the boundary in Figure 10.6 loses all coherency, then the configuration shown in Figure 10.7*a* obtains. This particular configuration is formally equivalent to that of the torn tilt boundary shown in Figure 6.1*b*. In particular, there is no interaction between the two grains as is exemplified by the total lack of mutual distortion between the contacting faces. The individual grains may therefore be treated separately, independent of one another. Separate Burgers circuits can be associated with each, similar to that done in Figure 6.1*b*. It might be argued at this point that since the two free surfaces in Figure 10.7*a* contact one another, they cannot, in fact, be free surfaces. However, in our more general definition of a free surface, since the surfaces do not interact with one another, their energies must be identical to that which would obtain in the separated condition. We have thus forged a very strong link between energy and geometry. This is even more dramatically shown by reference to Figure 10.7*b* which is the torn twist counterpart of Figure 6.1*c*. Unlike the previous two cases, there are no dislocations present within such a body, but only the presence of newly created free surfaces that are shown by the dotted closure failures. Again, these surfaces are, in fact, free since they do not interact with one another.

Figure 10.8*a* illustrates the twist boundary counterpart of the symmetric tilt boundary shown in Figure 6.1*d* (Marcinkowski and Dwarakadasa, 1973). On the other hand, Figure 10.8*b* shows the configuration of this boundary right at the interface where both faces are forced into a common or coincidence site lattice. This coherent region is similar in structure to the elastically strained region of Figure 10.1*b*. It should be pointed out here that the state shown in Figure 10.8 refers to full coherency. States of lesser coherency could also be described with the fully noncoherent state being the other extreme limit. It follows that the crystal lattice dislocations associated with each grain in Figure 10.8 can be added to give a net result which consists of a cross grid of screw dislocations each of which terminates on an edge dislocation, as shown in Figure 10.8*b*.

When the twist boundary is generated by a pure plastic rotation, the configuration shown in Figure 10.9*a* obtains where the resemblance to its tilt boundary counterpart in Figure 6.3 is readily apparent. Just at the interface, where both faces of the crystal are strained to a common lattice, the configuration shown in Figure 10.9*b* obtains. Again, the grain boundary array consists of an orthogonal configuration of screw type dislocations. Because of its smaller size, the number of screw dislocations associated with the twist boundary of Figure 10.9 is less than that associated with Figure 10.8.

Similar to Figure 6.1*a*, we denote the state shown in Figure 10.6*a* as (k^{I1}). It can be generated from the (K) state by means of the distortion

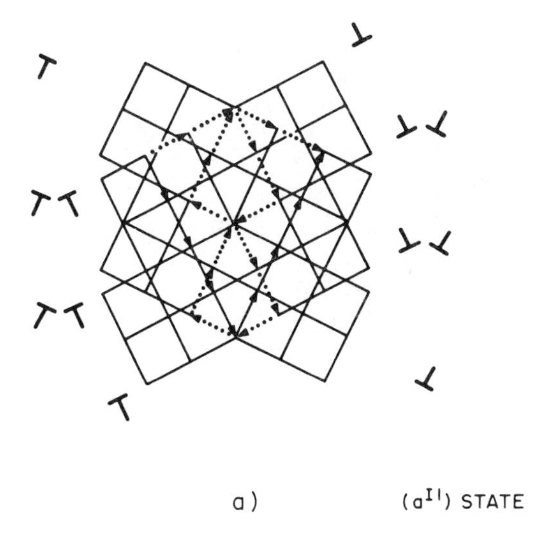

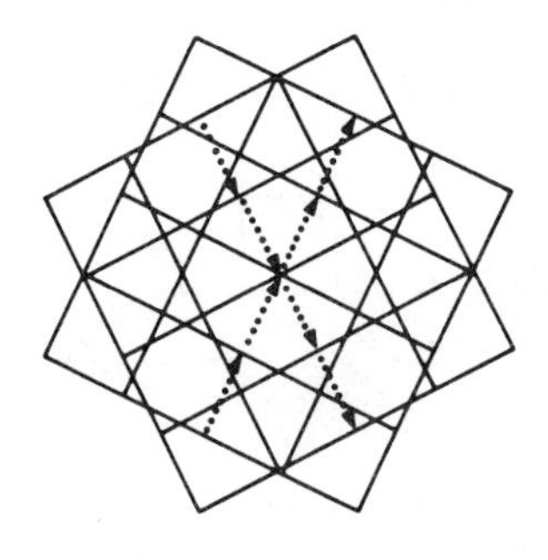

Figure 10.7 *a*) Twist boundary counterpart of symmetric boundary shown in Figure 6.1*b*. *b*) Torn twist counterpart of Figure 6.1*c*.

tensor $A_K^{k'1}$ whose nonvanishing components are

$$A_1^1 = A_2^2 = A_3^3 = 1 \tag{10.12a}$$

and

$$A_2^1 = \{H(-x^3)\tan\theta\}_1 + \{-H(+x^3)\tan\theta\}_2 \tag{10.12b}$$

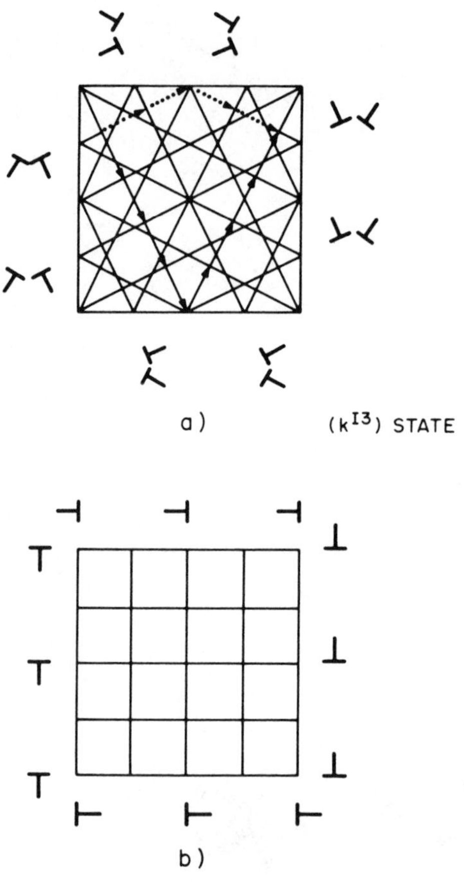

Figure 10.8 *a*) Twist boundary counterpart of symmetric tilt boundary shown in Figure 6.1*d*. *b*) Configuration right at the boundary.

that yields for the torsion tensor

$$S_{23}^{\cdot\cdot\,1} = -\tfrac{1}{2}\bar{A}_2^2\bar{A}_3^3\partial_3 A_2^1 = \left\{-\left(\tfrac{1}{2}\right)\delta(x^3)\tan\theta\right\}_1 + \left\{-\left(\tfrac{1}{2}\right)\delta(x^3)\tan\theta\right\}_2 \tag{10.13}$$

that when substituted into Eq. 6.4 yields the same result as that given by Eq. 6.6. On the other hand, when Eq. 10.13 is substituted into Eq. 6.7, we obtain for the dislocation density

$$\alpha^{11} = -2S_{23}^{\cdot\cdot\,1} = \left\{\delta(x^3)\tan\theta\right\}_1 + \left\{\delta(x^3)\tan\theta\right\}_2 \tag{10.14}$$

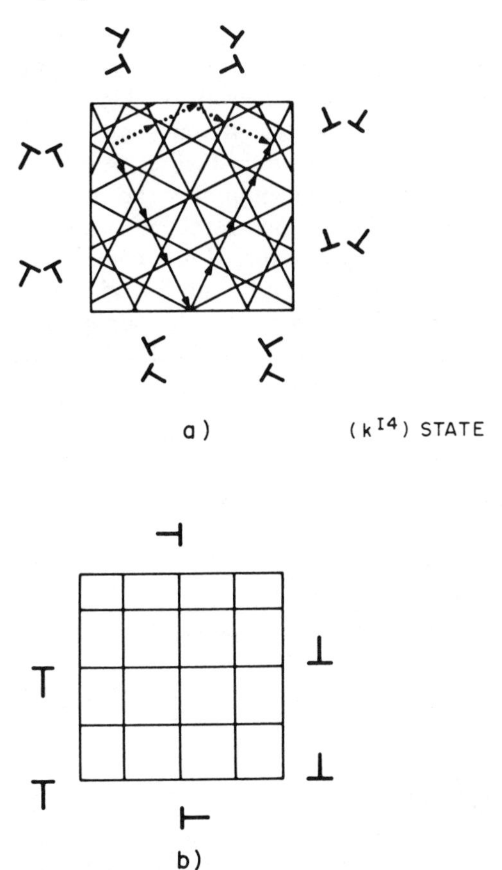

Figure 10.9 *a*) Twist boundary counterpart of the symmetric tilt boundary shown in Figure 6.3. *b*) Configuration right at boundary.

which is similar to that given by Eq. 6.8. In the case of the (a^{I1}) state shown in Figure 10.7*a*, the distortions are again the same as that given by Eq. 6, but with x^1 replaced by x^3. Also, analogous to Eq. 6.15, we obtain

$$S_{23}^{\cdot\cdot 1} = \Omega_{23}^{\cdot\cdot 1} \tag{10.15}$$

so that this component of the anholonomic object just compensates the torsion tensor. In addition, there is a nonvanishing component $\Omega_{23}^{\cdot\cdot 2}$ that accounts for the nondislocation related creation of free surface, analogous to that shown for the (a^{I2}) state of Figure 10.7*b*. The distortions that generate the (k^{I3}) and (k^{I4}) states of Figures 10.8*a* and 10.9*a*, respectively, from the (K) state are similar to those given by Eqs. 6.21 and 6.24, respectively, after replacing x^1 by x^3. The required tensor quantities can then be obtained quite readily from these distortions (Marcinkowski, 1976). Finally, in order to complete the analysis of twist boundaries, it follows that the configuration shown in Figure 10.3a is the twist boundary counterpart of the asymmetric tilt boundary shown in Figure 6.7c. However, in order to generate dislocations at this boundary as shown, the faces common to the boundary must be forced into coincidence.

The surface dislocation arrays and grain boundaries considered in this section can also be treated from a linearized point of view by utilizing the methods of Section 9. Consider, for example, the (k^{I1}) state twist boundary of Figure 10.6a. In their linearized form Eq. 10.12 can be written as

$$\beta_{12}^{P}=\{H(-x_3)\theta\}_1+\{-H(+x_3)\theta\}_2 \qquad (10.16)$$

or alternately, along the same lines as that used to obtain Eq. 9.45, as

$$\beta_{12}^{P}=\{2H(-x_3)\theta\}_1=\{-2H(+x_3)\theta\}_2 \qquad (10.17)$$

from which we obtain the following torsion tensor:

$$S_{231}=\partial_{[3}\beta_{1]2}^{P}=\left\{-\left(\tfrac{1}{2}\right)\theta\delta(x_3)\right\}_1+\left\{-\left(\tfrac{1}{2}\right)\theta\delta(x_3)\right\}_2\equiv\{-\theta\delta(x_3)\} \qquad (10.18)$$

which can be integrated to yield the following closure failure:

$$b_1=-\int_S S_{231}\,dF_{23}=2\theta\Delta x_2=8\theta \qquad (10.19)$$

Similar procedures can be followed for linearized versions of all the remaining dislocation arrays considered in this section.

REVIEW

When the Burgers vector lies parallel to a dislocation line, the dislocation is said to be of pure screw type. Such screw-type arrays have been the subject of the present chapter. In particular, arrays of internal or external screw-type arrays can be analyzed in a manner similar to that done for edge-type arrays in Chapters 4 and 5, respectively. When the internal arrays consist of a cross grid of screw-type dislocations, a twist-type grain boundary is said to be the result. Such boundaries may be treated in a manner similar to that used for tilt-type grain boundaries considered in Chapter 6. The Burgers circuits for such arrays are somewhat more complex than those for edge arrays since they must be carried out in three-dimensional space, as illustrated in Figure 10.5. On the other hand, the relevant tensor quantities such as Burgers vector, distortion, torsion and dislocation density are still of the same form as those obtained for edge dislocation arrays. In the case of the dislocation density tensor α^{kl}, $k=l$ for the screw-type arrays, as contrasted with $k \neq l$ for edge-type arrays. The completion of this chapter now enables us to completely describe the dislocation configuration of any arbitrarily oriented grain boundary. In addition, any state of internal stress can be described in terms of some suitable surface dislocation array comprised of both edge and screw-types.

CHAPTER 11

Relationship Between Stress and Strain in the Linear Approximation

Our treatment of dislocations until now has been concerned only with their geometrical aspects. Before we can apply these results to practical problems, it is first necessary to obtain equations that describe the stress fields associated with various dislocation arrays. Again, the only significant progress in this area has been in the limit of small distortions where the relationship between stresses and strain is taken to be linear. In spite of this limitation, the results obtained with this assumption are seen to be of great importance in understanding most of the behavior associated with the mechanical behavior of matter.

Since the elastic and plastic strains given by Eqs. 8.8 and 8.40 are formally equivalent to one another, it follows that a pure plastic strain should also satisfy the compatibility conditions, analogous to that given by Eq. 8.54 for a pure elastic strain. In particular, we can write in the linear approximation

$$4(\partial_{ab}e_{cd}^{P})_{[ad][bc]} \equiv \partial_{ab}e_{cd} - \partial_{db}e_{ca} + \partial_{dc}e_{bd} - \partial_{ac}e_{bd} = 0 \tag{11.1}$$

It is also clear from the way that the plastic strain is given by Eq. 3.13 that we could also write an expression for the curvature tensor R_{adbc} analogous to that given by Eq. 8.57 for the elastic strain, except that the metric tensor would have to be given by

$$g_{KL} = a_{ab}A_K^a A_L^b = \delta_{ab}A_K^a A_L^b \tag{11.2}$$

that is, the metric for the (a) state expressed in (K) state coordinates. Note that the metric used in Eq. 8.57 was indeed that associated with the deformed (κ) state. Thus we have similar to Eq. 8.58

$$4(\partial_{ab}e_{cd})_{[ad][bc]}=R_{adbc} \tag{11.3}$$

When we consider the (k) type states such as the (k') state of Figure 4.2, it is apparent that the compatibility conditions become much more complicated. In particular, we now have both elastic plus plastic strains present so that according to Eq. 8.50

$$e_{kl}^{T}=e_{kl}+e_{kl}^{P} \tag{11.4}$$

It also follows that the above equation could also be written as

$$e_{kl}^{T}=\tfrac{1}{2}\left(\partial_k u_l^T+\partial_l u_k^T\right) \tag{11.5}$$

Let us now consider the specific case of the (k') state shown in Figure 4.2. The plastic strain can be written as

$$e_{22}^{P}=\beta_{22}^{P}=\{(V-1)H(-x_1)\}_1 \tag{11.6}$$

where β_{22}^{P} is given by Eq. 9.3. The total strain, on the other hand, may be obtained by writing the total displacement as

$$u_2^T=x_2(V-1)\left(\frac{1}{\pi}\right)\left[-\tan^{-1}\left(\frac{x_1}{x_2}\right)\right]+\left(\frac{1}{2}\right)x_2(V-1) \tag{11.7}$$

which reduces to zero for $x_2=0$, that is, the center of Figure 4.2. It also becomes zero for $x_1\to\infty$ and becomes $x_2(V-1)$ for $x_1\to-\infty$, as is the case in Figure 4.2. From Eq. 11.5, the components of strain are found to be

$$e_{22}^{T}=\partial_2 u_2^T=(V-1)\left[\left(\frac{1}{\pi}\right)\frac{x_1x_2}{(x_1^2+x_2^2)}-\left(\frac{1}{\pi}\right)\tan^{-1}\left(\frac{x_1}{x_2}\right)+\left(\frac{1}{2}\right)\right] \tag{11.8a}$$

and

$$e_{12}^{T}=e_{21}^{T}=\frac{1}{2}\,\partial_1 u_2^T=-\left(\frac{1}{2\pi}\right)(V-1)\frac{x_2^2}{(x_1^2+x_2^2)} \tag{11.8b}$$

In view of Eq. 11.4, the elastic strain is simply

$$e_{22}=e_{22}^{T}-e_{22}^{P} \tag{11.9a}$$

and

$$e_{12}=e_{21}=e_{12}^{T} \tag{11.9b}$$

Similar to that given by Eqs. 8.54 and 11.1, the compatibility conditions for the total strain can be written as

$$4(\partial_{mr}e_{kl}^{T})_{[ml][rk]}\equiv\partial_{mr}e_{kl}^{T}-\partial_{lr}e_{km}^{T}+d_{lk}e_{rm}^{T}-\partial_{mk}e_{rl}^{T}=0 \tag{11.10}$$

while

$$4(\partial_{mr}e_{kl})_{[ml][rk]}=R_{mlrk} \tag{11.11}$$

It can be verified by use of Eq. 11.8 that Eq. 11.10 holds, that is, the compatibility conditions for the total strain are satisfied. Now, however, when we write the compatibility equations for the plastic strain we obtain

$$\partial_{mr}e_{kl}^{P}=\partial_{lr}e_{km}^{P}+\partial_{lk}e_{rm}^{P}-\partial_{mk}e_{rl}^{P}=G_{mlrk} \tag{11.12}$$

The compatibility conditions thus are not satisfied and G_{mlrk} is termed the incompatibility tensor. This can be easily shown for the specific case of the (k') state since from Eqs. 11.6 and 11.2 we obtain

$$G_{1122}=\partial_{11}e_{22}=-(V-1)\partial_{1}\delta(x_{1})=(V-1)\delta(x_{1})/x_{1} \tag{11.13}$$

where use has been made of the relation (deWit, 1973)

$$x\delta(x)=0 \tag{11.14}$$

which upon differentiating becomes

$$x\delta'(x)+\delta(x)=0 \tag{11.15}$$

It follows from Eq. 11.4 that the elastic strains e_{kl} separately also do not satisfy the compatibility conditions. This, however, does not imply that the presence of both elastic and plastic distortions will always imply incompatibility. For example, we may superimpose a compatible elastic strain on a compatible plastic strain. This could be done by superimposing the (κ) state distortions shown in Figure 2.1*b* on the (a) state distortion shown in Figure 3.1*a*. It also follows from Eqs. 11.4, 11.10, and 11.11 that

$$\partial_{mr}e_{kl}-\partial_{lr}e_{km}+\partial_{lk}e_{rm}-\partial_{mk}e_{rl}=-G_{mlrk} \tag{11.16}$$

Thus, while Eq. 11.12 states that dislocations are the source of the incompatibility, Eq. 11.16 states that the incompatibility is the source of the elastic strain. It also follows that dislocations need not be the only cause of the incompatibility, but can arise from any internal source of stress-free strain such as phase transformations, as for example the (κ^I) state of Figure 6.8*a*, localized thermal expansions, and so on.

Let us now assume that the stress and strain are related to one another by Hooke's law, that is (Kröner, 1958, deWit, 1960),

$$\sigma_{kl} = C_{klij} e_{ij} \tag{11.17}$$

where C_{klij} are the elastic constants. In the case of an elastically isotropic body, the above equation becomes

$$\sigma_{kl} = 2\mu\left(e_{kl} + \frac{1}{\nu - 2} e_{ii}\delta_{kl}\right) \tag{11.18a}$$

or

$$e_{kl} = \sigma_{kl} - \frac{1}{\nu + 1} \sigma_{ii}\delta_{kl} \tag{11.18b}$$

where μ and ν are the shear modulus and Poisson's ratio, respectively. Now Eq. 11.16 can be rewritten as

$$4(\partial_{mr} e_{kl})_{[ml][rl]} = -G_{mlrk} \tag{11.19}$$

where the square bracket notation shows that the indices are antisymmetric in *ml* and *rl*. We can thus define a new tensor η_{pq} as follows:

$$\eta_{pq} = \tfrac{1}{4}\varepsilon_{pml}\varepsilon_{qrk} G_{mlrk} \tag{11.20}$$

and rewrite Eq. 11.19 as

$$\varepsilon_{pml}\varepsilon_{qrk}\partial_{mr} e_{kl} = -\eta_{pq} \tag{11.21}$$

where it is noted that the new incompatibility tensor η_{pq} is symmetric. We can now substitute Eq. 11.18b into Eq. 11.21 which together with the following equation of equilibrium:

$$\partial_i \sigma_{kl} = 0 \tag{11.22}$$

yields

$$\partial_{kk}\sigma_{ij} + \frac{\nu}{\nu+1}(\partial_{ij}\sigma_{kk} - \partial_{ll}\sigma_{kk}\delta_{ij}) = 2\mu\eta_{ij} \tag{11.23}$$

The problem of internal stress now becomes one of simultaneously solving the last two field equations. We can next write the stress in terms of the incompatibility of the stress function ψ_{jn} as follows:

$$\sigma_{kl} = -\varepsilon_{kij}\varepsilon_{lmn}\partial_{im}\psi_{jn} \tag{11.24}$$

where the stress function can alternately be written in terms of a different stress function χ_{kl}

$$\psi_{kl} = 2\mu\left(\chi_{kl} + \frac{1}{\nu - 1}\chi_{mm}\delta_{kl}\right) \tag{11.25}$$

The last two equations can also be used to find

$$\partial_{kkll}\chi_{ij} = \eta_{ij} \tag{11.26}$$

The solution of this equation is given by

$$\chi_{ij} = -\frac{1}{8\pi}\int \eta_{ij}(\mathbf{r}')|\mathbf{r} - \mathbf{r}'|\,dV' \tag{11.27}$$

where $-1/8\pi|\mathbf{r}-\mathbf{r}'|$ is the Green's function for ∂_{kkll} in the whole of space. In particular, the Green's function corresponds to a displacement at $\mathbf{r}$ due to a point force at $\mathbf{r}'$. Now for the specific case of dislocations, we can write

$$\eta_{ij} = -\left(\varepsilon_{jmn}\partial_m\alpha_{in}\right)^{(s)} \tag{11.28}$$

where (s) denotes that only the symmetric part of the tensor is to be considered. Equations 11.28, 11.27, and 11.25 can then be used to obtain the components of stress σ_{kl} given by Eq. 11.24 associated with a dislocation. In particular, for a straight screw dislocation lying along the z direction, the two nonvanishing components of stress become

$$\sigma_{zx} = -\frac{\mu b}{2\pi}\frac{y}{x^2 + y^2} \tag{11.29a}$$

and

$$\sigma_{zy} = \frac{\mu b}{2\pi}\frac{x}{x^2 + y^2} \tag{11.29b}$$

whereas for an edge dislocation with Burgers vector along x and lying along z, we obtain

$$\sigma_{xx} = -\frac{\mu b}{2\pi(1-\nu)}\frac{y(3x^2+y^2)}{(x^2+y^2)^2} \tag{11.30a}$$

$$\sigma_{yy} = \frac{\mu b}{2\pi(1-\nu)}\frac{y(x^2-y^2)}{(x^2+y^2)^2} \tag{11.30b}$$

$$\sigma_{xy} = \frac{\mu b}{2\pi(1-\nu)}\frac{x(x^2-y^2)}{(x^2+y^2)^2} \tag{11.30c}$$

$$\sigma_{zz} = \nu(\sigma_{xx} + \sigma_{yy}) = -\frac{\mu b\nu}{\pi(1-\nu)}\frac{y}{(x^2+y^2)} \tag{11.30d}$$

Thus far Eq. 11.27 has been used to solve the problem of a single dislocation. For more complicated dislocation arrays or for more generalized states of internal stress, the solution to this problem becomes extremely complicated if not impossible. However, we have seen in previous sections that any state of stress either external or internal can be represented by some suitable array of dislocations. The problem thus becomes one of formally treating such arrays. We now see that this becomes possible by employing the techniques of continuous distributions of dislocations (Eshelby, 1956; Lardner, 1974).

In order to demonstrate the power of the continuous distribution dislocation method, let us consider the screw dislocation parallel to the surface of a semi-infinite or half space such as shown in Figure 11.1*a* (Jagannadham and Marcinkowski, 1978b). In order that the surface tractions from this dislocation vanish on the surface of the crystal, it is only necessary to distribute a continuous set of infinitesimal screw-type surface dislocations of opposite sign on this surface as described in Section 5 and as is also shown in the figure. Let $b_s f(y')dy'$ be the amount of Burgers vector lying between y' and $y'+dy'$. Then from Eq. 11.29a the value of σ_{xz} at some point y due to all the other dislocations on this surface, except that in the neighborhood of y, is given by

$$\int_{-\infty}^{y-\varepsilon} + \int_{y+\varepsilon}^{\infty} \frac{\mu b_s}{2\pi}\frac{f(y')\,dy'}{(y-y')} \tag{11.31}$$

Taking the limit as $\varepsilon \to 0$, we obtain the Cauchy principal value

$$⨍_{-\infty}^{+\infty} \frac{\mu b_s}{2\pi}\frac{f(y')\,dy'}{(y-y')} \tag{11.32}$$

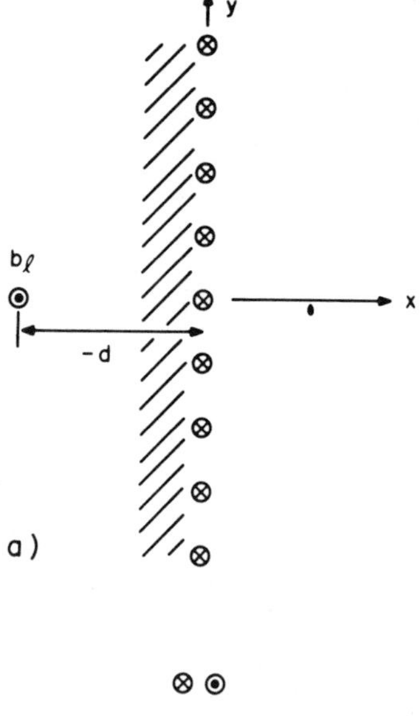

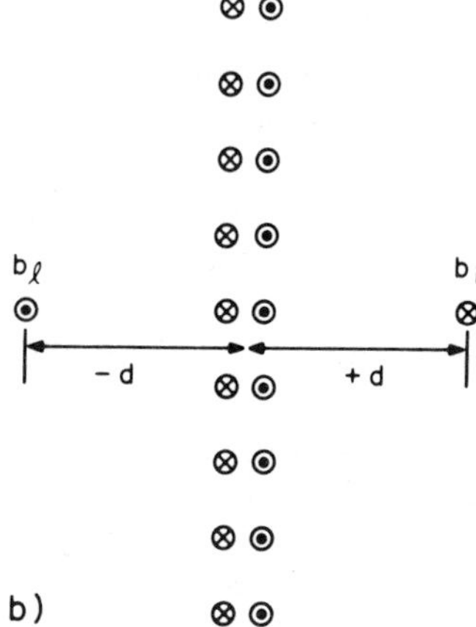

Figure 11.1 *a*) Screw-type dislocation near a free surface. *b*) Screw-type dislocation dipole. Reprinted by permission of Internationaler Buch-Versand GmbH from K. Jagannadham and M. J. Marcinkowski, (1978b) *Physica Status Solidi*, **a50**, Figures 4 and 2.

The condition that the component of stress σ_{xz} vanish on the surface of the crystal at y is simply

$$\int_{-\infty}^{+\infty} \frac{\mu b_s}{2\pi} \frac{f(y')\,dy'}{(y-y')} - \frac{\mu b_l}{2\pi} \frac{y}{(d^2+y^2)} = 0 \tag{11.33}$$

where the second term is the stress at the surface due to the crystal lattice dislocation. The solution, that is, inversion of the above integral yields

(Jagannadham and Marcinkowski, 1978b)

$$f(y)=\frac{b_l}{\pi b_s}\frac{d}{(d^2+y^2)} \tag{11.34}$$

Thus, the density of dislocations as measured by the distribution function $f(y)$, vanishes as d or y go to infinity, or as $d=0$. The stress due to the surface dislocation array can be found from Eqs. 11.34 and 11.29a to be

$$\sigma_{xz}^{s}=\frac{\mu b_l d}{2\pi^2}\int_{-\infty}^{+\infty}\frac{(y-y')\,dy'}{(d^2+y'^2)[x^2+(y-y')^2]} \tag{11.35}$$

which yields

$$\sigma_{xz}^{s}=\frac{\mu b_l}{4\pi}\left[\frac{y}{(x+d)^2+y^2}+\frac{y}{(x-d)^2+y^2}-\frac{4xyd}{\{(x+d)^2+y^2\}\{(x-d)^2+y^2\}}\right] \quad \text{for } x>0 \tag{11.36a}$$

whereas

$$\sigma_{xz}^{s}=\frac{\mu b_l}{4\pi}\left[\frac{y}{(x+d)^2+y^2}+\frac{y}{(x-d)^2+y^2}+\frac{4xyd}{\{(x+d)^2+y^2\}\{(x-d)^2+y^2\}}\right] \quad \text{for } x<0 \tag{11.36b}$$

Since the stress field of the crystal lattice dislocation is

$$\sigma_{xz}^{l}=-\frac{\mu b_l}{2\pi}\frac{y}{(x+d)^2+y^2} \tag{11.37}$$

addition of the last two equations yields the total stress field or

$$\sigma_{xz}^{T}=0 \quad \text{for } x>0 \tag{11.38a}$$

and

$$\sigma_{xz}^{T}=\frac{2\mu b_l}{\pi}\frac{dxy}{[(x+d)^2+y^2][(x-d)^2+y^2]} \quad \text{for } x<0 \tag{11.38b}$$

Thus, the vertical array of surface dislocations in Figure 11.1*a* completely screens the stress field of the crystal lattice dislocation from the vacuum, which must be the condition physically. The configuration shown in this figure could also be viewed as occurring in an infinite medium in which case the rightmost portion of the body could be cut from the leftmost portion since it would be stress free. It has been argued that such free-surface problems could be treated by a method similar to that used in electrostatics in which an image dislocation b_i of opposite sign to that of the crystal lattice dislocation b_l is placed in the vacuum at an equal distance from the surface as that of the crystal dislocation such as shown in Figure 11.1*b* (Head, 1953). This argument, however, is physically wrong since it is meaningless to speak of a dislocation in a vacuum. On the other hand, it is meaningful to speak of an electrostatic charge associated with an image charge in a vacuum. The physical interpretation of what one really does in employing the so-called image dislocation method is shown in Figure 11.1*b*. In particular, a dislocation dipole is placed in an infinite medium. The surface tractions are seen to vanish along a plane midway between the dislocations comprising the dipole. One then thinks that he should be able to make a cut along this plane followed by separation of the two half crystals. However, this is not possible unless a set of surface dislocations is placed on each of the cut surfaces in the manner described in Figure 11.1*a*. Upon rejoining with one another, the two surface arrays that are shown in Figure 11.1*b* annihilate with one another. Not only is the image dislocation method difficult to employ for more complex dislocation configurations because of the need of multiple images, but it breaks down completely for dislocations in arbitrary shaped bodies of finite dimensions. No such handicap exists in the method of surface dislocations.

It is also apparent that the total Burgers vector associated with all of the surface dislocations in Figure 11.1 can be written as

$$\int_{-\infty}^{+\infty} b_s f(y)\,dy = b_l \tag{11.39}$$

where $f(y)$ given by Eq. 11.34 has been employed. This is simply another manifestation of the conservation rule given by Eq. 5.80. It simply states that there can never be a dislocation with Burgers vector b_l inside an infinite body without somewhere finding negative counterparts whose sum adds up to the magnitude of b_l. There are still other corollaries of this very powerful principle which says that dislocations must always be created in pairs. This seems to be connected with other basic concepts in physics such as that which states that every particle must have associated with it an antiparticle or all matter must possess an equal amount of antimatter.

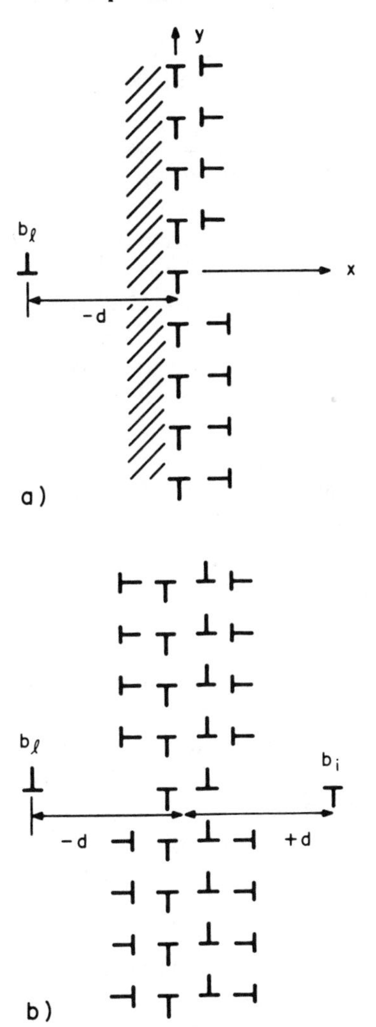

Figure 11.2 *a*) Edge-type dislocation near a free surface. *b*) Edge-type dislocation dipole. Reprinted by permission of Internationaler Buch-Versand GmbH from K. Jagannadham and M. J. Marcinkowski, (1978b), *Physica Status Solidi*, **a50**, Figures 5 and 3.

For completeness, Figure 11.2 shows the surface dislocation array associated with an eedge dislocation near a free surface. In this case, two distinct surface arrays are required, one with Burgers vector perpendicular to the surface, that is, the primary array, and one with Burgers vector parallel to the surface, that is, the secondary array. The distribution functions and stresses associated with these two arrays can be found using the continuous distribution methods already discussed for the screw type dislocation. Here it is found that the sum of the Burgers vectors of the crystal lattice dislocations and primary array of surface dislocations is zero, whereas that of the secondary array itself is zero. This latter result follows from the fact that the dislocations in the secondary array occur in pairs of opposite sign. If next we combine two half spaces in the manner employed for the screw dislocation the configuration shown in Figure 11.2*b* obtains. It shows that whereas the primary arrays from the two half spaces cancel with one another, those associated with the secondary array combine and thus remain at the interface. This shows that not only is the image dislocation b_i necessary to satisfy the free surface boundary conditions, but a secondary dislocation array as well. This proves once again that the image dislocation concept is not in general correct and that the surface dislocation model suffers from no such deficiencies. The surface dislocation method is also readily applicable to the somewhat more generalized case in which the vacuum in Figures 11.1*a* and 11.2*a* is replaced by a second phase with elastic constants different from that of the phase towards the left (Jagannadham and Marcinkowski, 1978d).

REVIEW

Hooke's law, as given by Eq. 11.17, forms the basis by which stress and strain are related to one another in the linear approximation. An internal stress, say as arises from a dislocation, can in turn be written in terms of the incompatability of a stress frunction ψ_{jn}, as embodied in Eq. 11.24. For single infinitely long screw and edge-type dislocations lying along the z axis, the stress fields are given by Eqs. 11.29 and 11.30, respectively. It is important to note that singularities in the stress occur at the dislocation core, that is, $x=0$ and $y=0$.

The stress given by Eqs. 11.29 and 11.30 obviously refer to quantized dislocations. Consider what happens when we smear the dislocations out in space, say along a plane, such that they are infinite in number with each possessing a Burgers vector of infinitesmal strength. We may describe such arrays in terms of a distribution function $f(y)$. It is then possible to form an integral that yields the stress at some point due to all the other dislocations in the distribution except those just at the point in question. This procedure simply defines the method by which the Cauchy principal value of an integral is obtained and is given by Eq. 11.32.

Consider now the specific case of a single quantized screw-type dislocation within a semi-infinite crystal as depicted in Figure 11.1*a*. In order to satisfy the stress-free boundary conditions, it is necessary to place a continuous distribution of surface dislocations on the surface of the body, also shown in Figure 11.1*a*. The distribution function $f(y)$ associated with this array is given by Eq. 11.34 that may in turn be used to find the corresponding components of stress associated with this array, as given by Eq. 11.36. We note that, in contrast to those results obtained for a single dislocation, there is no singularity associated with the continuous distribution, that is, $\sigma_{kl} \neq \infty$ at $x=0$ and $y=0$. This is an important result and arises from the fact that in taking the Cauchy principal value of the integral, we have omitted the insignificant stress contribution from each infinitesimal dislocation at that point. This is the very essence of the argument that allows us to use a continuous distribution of dislocations to describe any state of elastic strain, and not have to worry about any singularities due to the dislocations in the array. It is of course true that in numerical calculations a finite number of discrete dislocations need to be employed. This in turn means that the results, as contrasted to those obtained for continuous arrays, are only approximate. However, the procedure can be made more accurate by increasing the number of dislocations, with a corresponding decrease in their Burgers vectors. It also follows that the total number of surface dislocations is equal in strength but opposite in sign to that of the single quantized crystal lattice dislocation, as given by Eq. 11.39.

In conclusion, it should also be pointed out that the so-called image dislocation model, as depicted in Figure 11.1*b*, has frequently been used to satisfy the stress-free boundary conditions. In this case, a fictitious image dislocation is placed in a vacuum at a distance from the surface equal to that of the real dislocation. It is shown that this idea is conceptually erroneous and only accidently leads to what appear to be correct results in the case of the screw-type dislocation. On the other hand, this method breaks down completely fro the edge-type dislocation. No such conceptual difficulties are encountered in the use of surface dislocations.

CHAPTER 12

Disclinations

The purpose of this chapter is to treat what has come to be called a whole new class of defects, namely the disclination. However, as closer scrutiny shows, such defects may be described in terms of special arrays of dislocations. For example, a grain boundary that terminates abruptly within a crystal, may be viewed as a wedge-type disclination. It has been argued that such defects possess too high an energy to be present in metals. On the other hand, in materials where the elastic constants are small, such as in certain biological organisms, these defects may be rather common. Also, and equally important, it is shown that disclinations yield a heuristic approach to the study of defects which enable much broader generalizations to be obtained with respect to the mechanical behavior of matter.

Let us consider the creation of a symmetric tilt boundary similar to that shown in Figure 6.1*a* but that now terminates within the interior of the crystal. The resulting configuration is shown in Figure 12.1*a* and may be termed a negative wedge disclination of strength $2\theta = 53.1°$ (Nabarro, 1967; Marcinkowski, 1977d). The angle 2θ is simply the angle of misorientation corresponding to that of the grain boundary in Figure 6.1*a*. Because of the high strain energy associated with the grain boundary of Figure 12.1*a*, it may be torn similar to that given by Figure 6.1*b* to yield the state shown in Figure 12.1*b*. This may be referred to as a partially torn state due to the presence of elastic strain that still remains after this imperfect tearing. All intermediate states between those shown in Figure 12.1*a* and 12.1*b* with configurations of the type shown in Figure 7.5 can also be obtained. The perfectly torn state corresponding to Figure 12.1*a*, in which all of the elastic distortions are removed, is shown in Figure 12.1*c*.

It is instructive at this point to consider a series of states similar to those shown in Figures 12.1*a–c*, but for a single dislocation. In particular, consider the dislocated state shown in Figure 5.11*a*. The very severe elastic

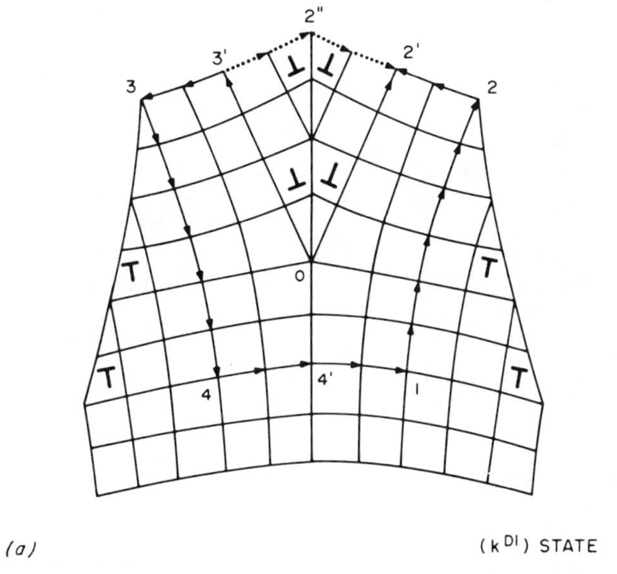

Figure 12.1a Wedge disclination of strength 53.1° without free surfaces.

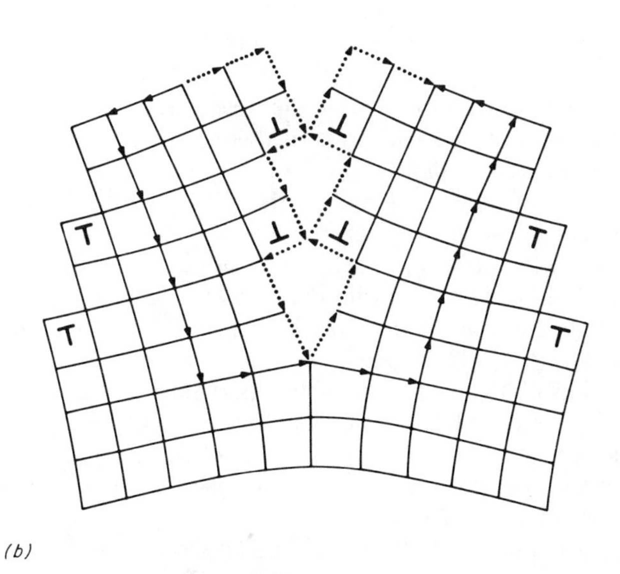

Figure 12.1b Imperfectly torn state associated with the wedge disclination of Figure 12.1*a*.

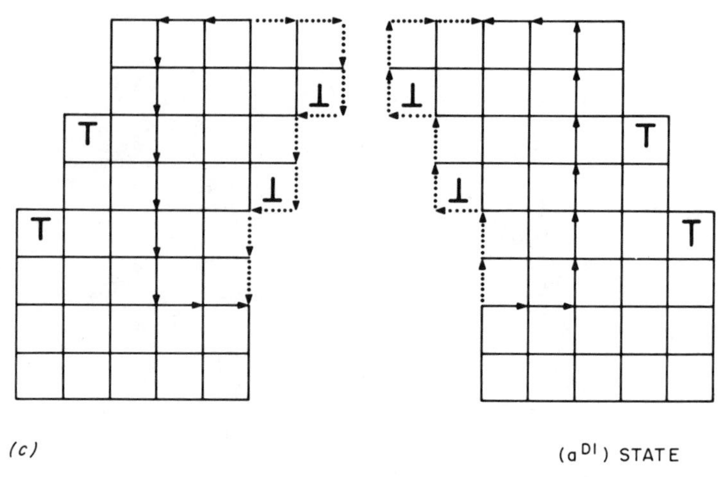

Figure 12.1c Perfectly torn state associated with the wedge disclination of Figure 12.1*a*.

distortions associated with this dislocation may be reduced by an imperfect tearing such as shown toward the left in Figure 12.2*a*. In reality, this imperfect tearing consists in the formation of an asymmetric tensile crack below the dislocation. The details of the fracture process are treated more fully in a subsequent section. The nature of the crack is shown in more detail toward the right in Figure 12.2. It is, in fact, seen to consist of an array of surface dislocations or dipoles, as discussed in Chapter 5. Note that the stress-free portions of the dipole, that have a free surface associated with them, cancel with the free surface associated with the crystal lattice dislocation. The stress field contribution thus arises from the dislocations situated at the bottom end of each dipole. Since the surface dislocations are continuous in nature, the partial tearing associated with a dislocation has allowed a previously quantized dislocation such as shown in Figure 5.11*a* to unquantize itself with a subsequent reduction in energy. The closure failure associated with the Burgers circuit of Figure 12.2*a* thus corresponds to the sum of the Burgers vectors of all the surface dislocations. Finally, Figure 12.2*b* illustrates the fully torn state associated with the dislocation of Figure 5.11*a*. In this particular case, all of the elastic distortions are removed. The disclinated state shown in Figure 12.1*b* may also be viewed as having continuous arrays of surface dislocations distributed over its torn surfaces in a manner similar to that shown for a single dislocation in Figure 12.2*a*.

Let us now consider the 90° grain boundary, similar to that illustrated in Figure 7.6*a*, but that ends within the crystal. The resulting configuration is

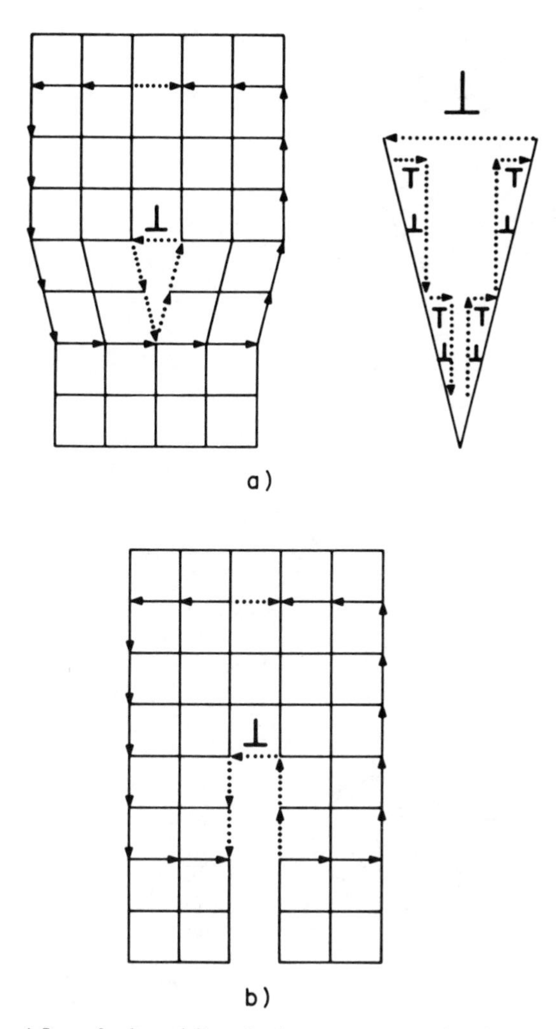

Figure 12.2 *a*) Imperfectly and *b*) perfectly torn state associated with Figure 5.11*a*.

shown in Figure 12.3*a* and is termed a 90° wedge disclination, and analogous to Figure 12.1*b*, is seen to be an imperfectly torn state. If now the voids in Figure 12.3*a* are filled with atoms, similar to the case in Figure 7.6*b*, the configuration in Figure 12.3*b* obtains. The insertion of these extra atoms is equivalent to uniformly distributing the dislocations in a continuous manner along the dotted line in Figure 12.3*b*. In this manner, the crystal lattice disclinations with quantized Burgers vectors have become

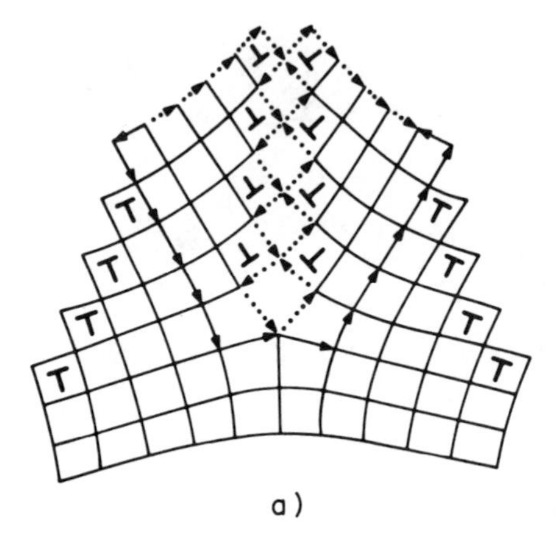

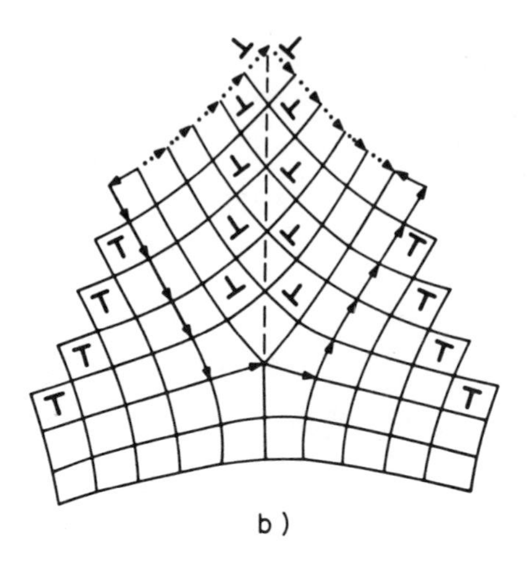

Figure 12.3 *a*) Imperfectly torn state associated with a wedge disclination of strength 90°. *b*) Perfect wedge disclination of strength 90°.

unquantized into a continuous distribution in Figure 12.3*b*. Disclinations of still greater strength can be obtained by involving distortions of the type shown in Figure 12.4. In particular this represents a perfect 180° wedge disclination and may be thought of as formed by the generation of an infinite number of edge-type dislocations on a single slip plane in two adjacent regions of a body. In general, the 90° and 180° wedge disclinations are drawn as shown in Figures 12.5*a* and 12.5*b*, respectively. In the former case, the disclination is seen to possess five-fold symmetry, whereas in the latter case six-fold symmetry is apparent. These may be termed perfect disclinations since they have no fault, that is, grain boundary associated with them. On the other hand, the 53.1° disclination shown in

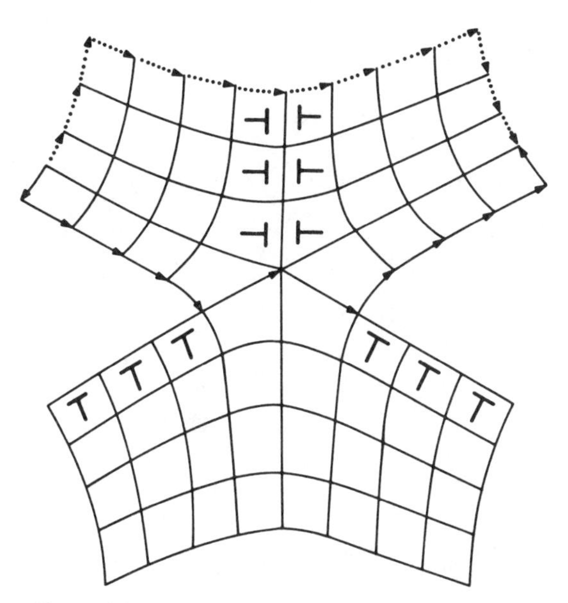

Figure 12.4 Perfect wedge disclination of strength 180°.

Figure 12.1 is an imperfect or partial disclination since it has a grain boundary attached to it. One might argue that the 90° and 180° disclinations have attached to them grain boundaries of 90° and 180° misorientation; however, such boundaries are perfect and thus possess no energy per unit length. An important question that we must now ask ourselves is whether the disclination is a basic defect independent of a dislocation or is it simply a special arrangement of dislocations. The answer appears to be the latter. For example, the 90° disclination in Figure 12.3 could move by first giving off an interstitial atom at its core followed by the generation of a pair of lattice dislocations at this same place. It is clear, however, that in the case of the 90° and 180° disclinations, the distortions are so large that nonlinear elasticity theory must be employed. Also obvious is the fact that the distortions associated with disclinations of high strength are so large as to rule out their presence in solids where the elastic constants are comparatively high. On the other hand, their occurrence in biological organisms seems well established.

As in the grain boundary of Figure 6.1*a*, a distortion tensor $A_K^{k^{D1}}$ can be associated with the disclination of Figure 12.1*a* that is denoted as the (k^{D1}) state. In particular, its nonvanishing components are

$$A_1^1 = A_2^2 = A_3^3 = 1 \tag{12.1a}$$

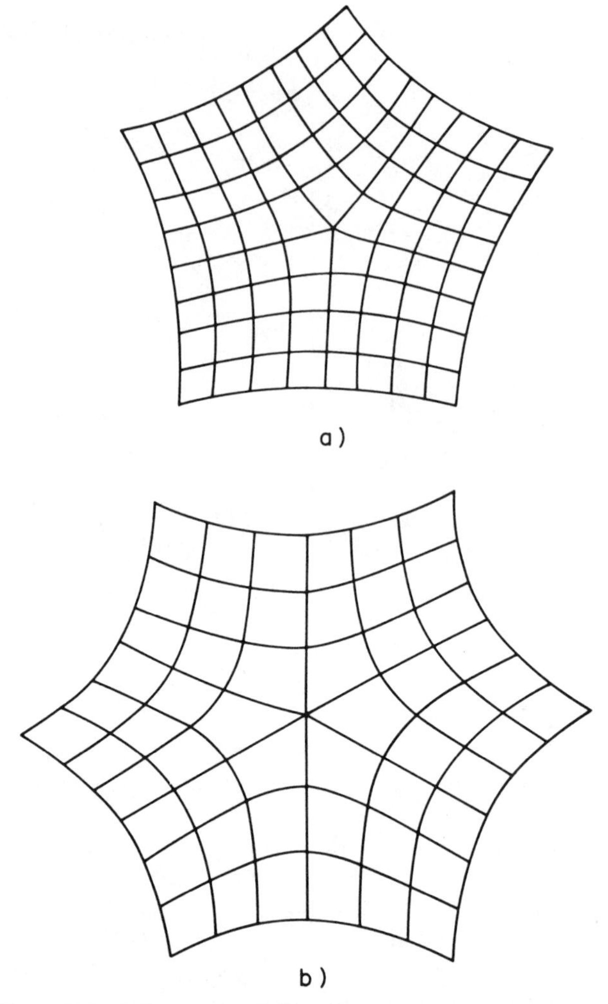

Figure 12.5 Perfect wedge disclinations of strength. *a*) 90°. *b*) 180°.

and

$$A_2^1 = \left\{\left(\frac{N}{M}\right)H(-x^1)H(+x^2)\right\}_1 + \left\{-\left(\frac{N}{M}\right)H(+x^1)H(+x^2)\right\}_2 \tag{12.1b}$$

where $\tan\theta$ in Eq. 6.1b has been replaced by (N/M) as in Eq. 7.2, since because of elastic strains the meaning of $\tan\theta$ is now lost. The distortion also contains the Heaviside function $H(+x^2)$ to indicate that the grain boundary terminates within the crystal. The Burgers vector can thus be written as

$$b^{k^{D1}} = -\oint A_K^{k^{D1}} dx^K \tag{12.2}$$

which gives

$$b^1 = -\int_3^4 A_2^1 dx^2 - \int_1^2 A_2^1 dx^2 \tag{12.3}$$

or

$$b^1 = \left\{4\left(\frac{N}{M}\right)\right\}_1 + \left\{4\left(\frac{N}{M}\right)\right\}_2 = \{2\}_1 + \{2\}_2 \tag{12.4}$$

which corresponds to the four dotted arrows associated with the Burgers circuit of Figure 12.1*a*. It is also apparent that an elastic distortion $B_K^{k^{D1}}$ is also associated with the disclination of Figure 12.1*a*, and if known, would allow the base vectors associated with the coordinates to be determined.

In place of the line integral of Eq. 12.2 we could have written

$$b^{k^{D1}} = -\int_s S_{l^{D1}m^{D1}}^{\cdot\cdot\;k^{D1}} dF^{l^{D1}m^{D1}} \tag{12.5}$$

The only nonvanishing component of $S_{l^{D1}m^{D1}}^{\cdot\cdot\;k^{D1}}$ can be found from Eqs. 12.1 and 4.102a to be

$$S_{12}^{\cdot\cdot\;1} = \left\{-\left(\frac{1}{2}\right)\delta(x^1)\left(\frac{N}{M}\right)H(+x^2)\right\}_1 + \left\{-\left(\frac{1}{2}\right)\delta(x^1)\left(\frac{N}{M}\right)H(+x^2)\right\}_2 \tag{12.6}$$

that when substituted into Eq. 12.5, again yields the same result as that given by Eq. 12.4. The dislocation density could also be determined by employing a relation similar to that given by Eq. 6.7 to yield

$$\alpha^{31} = -2S_{12}^{\cdot\cdot\;1} = \left\{\left(\frac{N}{M}\right)\delta(x^1)H(+x^2)\right\}_1 + \left\{\left(\frac{N}{M}\right)\delta(x^1)H(+x^2)\right\}_2 \tag{12.7a}$$

that in terms of Figure 12.1*a* is simply

$$\alpha^{31} = \left\{\frac{\underset{3'-2''}{\Delta x^1}}{\underset{0-3'}{\Delta x^2}}\right\}_1 + \left\{\frac{\underset{2''-2'}{\Delta x^1}}{\underset{0-2'}{\Delta x^2}}\right\}_2 \tag{12.7b}$$

Let us next write the expression for a quantity $R_{nml}^{\cdot\cdot\cdot k}$ termed the Riemann–Christoffel curvature tensor of the second kind given by (Anthony, 1970; Marcinkowski, 1977e)

$$R_{nml}^{\cdot\cdot\cdot k} = \partial_n \Gamma_{ml}^k - \partial_m \Gamma_{nl}^k + \Gamma_{nr}^k \Gamma_{ml}^r - \Gamma_{mr}^k \Gamma_{nl}^r \tag{12.8}$$

which is related to the curvature tensor of the first kind given by Eq. 8.55 as follows:

$$R_{mnli} = g_{ik} R_{mnl}^{\cdot\cdot\cdot k} \tag{12.9}$$

Since $R_{nml}^{\cdot\cdot\cdot k}$ is antisymmetric in the indices nm, we can write for the only nonvanishing components of this tensor

$$R_{122}^{\cdot\cdot\cdot 1} = -R_{212}^{\cdot\cdot\cdot 1} = -\partial_2 \Gamma_{12}^1 - \Gamma_{21}^1 \Gamma_{12}^1 \tag{12.10}$$

However, from Eq. 5.30 since $\Gamma_{12}^1 = 2S_{12}^{\cdot\cdot 1}$ and since $\Gamma_{21}^1 = 0$, the above equation reduces to

$$R_{122}^{\cdot\cdot\cdot 1} = -R_{212}^{\cdot\cdot\cdot 1} = -2\partial_2 S_{12}^{\cdot\cdot 1} \tag{12.11}$$

so that in view of Eq. 12.6, we may write

$$R_{122}^{\cdot\cdot\cdot 1} = -R_{212}^{\cdot\cdot\cdot 1} = \left\{\left(\frac{N}{M}\right)\delta(x^1)\delta(x^2)\right\}_1 + \left\{\left(\frac{N}{M}\right)\delta(x^1)\delta(x^2)\right\}_2 \tag{12.12}$$

The question that now arises concerns the physical meaning of the curvature tensor. It has been argued that the torsion tensor alone is insufficient to describe the Burgers vector and that a more general expression for Eq. 12.5 should be written as (Kröner, 1958)

$$b^k = -\int_s \left[S_{nm}^{\cdot\cdot\cdot k} - \frac{1}{2} R_{nml}^{\cdot\cdot\cdot k} C^l \right] dF^{nm} \tag{12.13}$$

We now show that Eq. 12.13 is misleading and that instead we should write

$$b^{k^{D1}} = \int_s \frac{1}{2} R_{n^{D1}m^{D1}l^{D1}}^{\cdot\cdot\cdot k^{D1}} C^{l^{D1}} dF^{n^{D1}m^{D1}} \tag{12.14}$$

where Eq. 12.14 is simply an alternate representation of Eq. 12.5. In the former case we are dealing with the closure failure associated with a given circuit or path, while in the latter we are dealing with the closure failure associated with the parallel displacement of a vector $C^{l^{D}1}$ about a given path. We have already seen how Eq. 12.5 gives the closure failure shown in Figure 12.1*a*. In the case of Eq. 12.14, we may write

$$b^1 = \frac{1}{2}\int_S R_{122}^{\cdot\cdot\cdot 1} C^2 dF^{12} \tag{12.15}$$

where C^2 is the vector 0–3′ in Figure 12.1*a* that is displaced in a parallel manner about a closed path until it becomes the vector 0–2′. The closed path may be viewed as shrunk to the origin 0. The failure of the vector C^2 to coincide with itself after such a displacement is simply the closure failure b^1. It is seen from Eq. 12.12 that Eq. 12.15 yields the same result as that given by Eq. 12.4. We then see that a disclination possesses both a torsion tensor and a curvature and that from Eq. 12.11 the curvature is the result of the torsion. The question that now arises is whether the same also holds true for a dislocation? This can be seen by first considering the expression $S_{12}^{\cdot\cdot 1}$ given by Eq. 4.93 for a single dislocation. In view of Eqs. 12.11 and 11.15 we may write

$$R_{122}^{\cdot\cdot\cdot 1} = -R_{212}^{\cdot\cdot\cdot 1} = \left(\frac{1}{x^2}\right)\delta(x^1)\delta(x^2) \tag{12.16}$$

When substituted into Eq. 12.15 this yields

$$b^1 = \int_S \delta(x^1)\delta(x^2)\, dx^1 dx^2 \tag{12.17}$$

where we have utilized the fact that $C^2 \equiv x^2$. The vector under consideration with respect to parallel displacement in Figure 4.4*b* is 0–3′ displaced to 0–2′ about a circuit shrunk to the origin. Note that unlike the cases for the disclination in Figure 12.1*a*, the magnitude of the closure failure is independent of the length C^2 of the vector. This arises from the presence of the factor $(1/C^2) \equiv (1/x^2)$ in Eq. 12.16. Thus, we see that the dislocation as well as the disclination possess both curvature and torsion. This should not be surprising fact since from Eq. 12.9, the curvature is a measure of the defect content, that is, incompatability associated with a given body. Similar to the method used in conjunction with Eq. 4.114, we can rewrite Eq. 12.14 in differential form to obtain

$$db^k = \frac{1}{2} R_{nml}^{\cdot\cdot\cdot k} C^l dF^{nm} = \frac{1}{2} R_{nml}^{\cdot\cdot\cdot k} C^l \varepsilon^{rnm} dF_r = \alpha^{rk} dF_r \tag{12.18}$$

where

$$\alpha^{rk} = \tfrac{1}{2}\varepsilon^{rnm} R_{nml}^{\cdot\cdot\cdot k} C^{l} \tag{12.19}$$

We thus have an alternate way to Eq. 4.117 of obtaining the dislocation density.

The analysis that we have used for Figure 12.1*a* could also be applied to Figure 12.1*b*, although in this case the various tensor quantities would apply to the surface dislocations spread over the various cut surfaces. On the other hand, in the case of the fully torn state shown in Figure 12.1*c* we would be able to write, similar to Eq. 6.14

$$b^{a^{D1}} = -\int_{S} \left(S_{c^{D1} b^{D1}}^{\cdot \ \cdot \ a^{D1}} - \Omega_{c^{D1} b^{D1}}^{\cdot \ \cdot \ a^{D1}} \right) dF^{c^{D1} b^{D1}} \tag{12.20}$$

where

$$\Omega_{12}^{\cdot\cdot 1} = S_{12}^{\cdot\cdot 1} \tag{12.21}$$

The quantity $\Omega_{12}^{\cdot\cdot 1}$, as always, corresponds to the newly created free surfaces associated with each of the dislocations in Figure 12.1*c*. Note that this contribution does not appear in the case of Figure 12.1*b* since these free surfaces are just compensated by the free surface contributions arising from the surface dislocation dipoles. This is more clearly seen by reference to the rightmost side of Figure 12.2*a* corresponding to a single dislocation. Analyses, similar to that made for the 53.1° wedge disclination can also be carried out for the 90° and 180° disclinations shown in Figures 12.3 and 12.4, respectively. In the former case, N/M for the distortion tensor of Eq. 12.1b becomes 1, whereas in the latter case it becomes ∞. This latter result means that Figure 12.4 is generated by the motion of an infinite number of dislocations on a single slip plane. We then have from Eqs. 12.6 and 12.12 that the 180° disclination has associated with it both an infinite torsion and curvature tensor. Physically this means that when a vector is moved by parallel displacement about a closed path, it becomes antiparallel to itself upon returning to its point of origin.

Our treatment of disclinations thus far has been quite general, that is nonlinear. However, for such large distortions the constitutive equations between stress and strain have not yet been developed. Such, however, is not the case for small distortions. Consider for example the distortion tensor given by Eq. 12.1*b*. Since for small distortions we can once again consider the coordinates to be Cartesian, we can replace (N/M) by $\tan\theta$ that in turn can be replaced by θ so that Eq. 12.1*b* becomes

$$A_{12} = \beta_{12}^{P} = \{\theta H(-x_1) H(+x_2)\}_{,1} + \{-\theta H(+x_1) H(+x_2)\}_{,2} \tag{12.22}$$

The linearized form of the torsion tensor is found from Eq. 12.6 to be

$$S_{121}=\{-(\tfrac{1}{2})\theta\delta(x_1)H(+x_2)\}_1 + \{-(\tfrac{1}{2})\theta\delta(x_1)H(+x_2)\}_2 \equiv \{-\theta\delta(x_1)H(+x_2)\} \quad (12.23)$$

which yields

$$b^1 = -\int_s S_{121}\,dF_{12} = 2\theta\Delta x_2 = 8\theta \quad (12.24)$$

It also follows that the two forms of the curvature tensor given by Eq. 12.9 become identical so that from Eq. 12.11

$$R_{1221} = -R_{2121} = -2\partial_2 S_{121} = \{\theta\delta(x_1)\delta(x_2)\}_1 + \{\theta\delta(x_1)\delta(x_2)\}_2 \equiv \{2\theta\delta(x_1)\delta(x_2)\} \quad (12.25)$$

that in view of Eq. 12.15 yields

$$b^1 = \tfrac{1}{2}\int_s R_{1221}C_2\,dF_{12} = 8\theta \quad (12.26)$$

Again, as in the case of a grain boundary, the plastic distortion β_{kl}^P associated with a disclination can be decomposed into a pure shear part and a rotation part, that is,

$$\beta_{kl}^P = \beta_{kl}^{Pe} + \beta_{kl}^{P\omega} \quad (12.27)$$

We can also define a bend-twist tensor κ_{mkl}^P as follows:

$$\kappa_{mkl}^P = \partial_m \beta_{kl}^{P\omega} \quad (12.28)$$

that in view of Eq. 12.22 is simply

$$\kappa_{112}^P = \left\{-\left(\frac{\theta}{2}\right)\delta(x_1)H(+x_2)\right\}_1 + \left\{-\left(\frac{\theta}{2}\right)\delta(x_1)H(+x_2)\right\}_2 \quad (12.29)$$

where it is to be noted that the bend-twist is antisymmetric in its last two indices. We can also define a disclination density given by

$$\Gamma_{nmkl} \equiv 2\partial_{[n}\kappa^P_{m]kl} \quad (12.30)$$

that in view of Eq. 12.29 is given by

$$\Gamma_{2112} = \{-\theta\delta(x_1)\delta(x_2)\}_1 + \{-\theta\delta(x_1)\delta(x_2)\}_2 \qquad (12.31)$$

We see that this is identical to the curvature tensor given by Eq. 12.25, so that the curvature tensor by itself may be viewed as giving the disclination density. Note that it vanishes everywhere except at the origin of the disclination, as it should. It may also be discerned from Eq. 9.51 that while a bend twist exists for a grain boundary, there is nevertheless no curvature tensor, that is, disclination density associated with it. On the other hand, to the extent that a grain boundary is comprised of discrete dislocations, a curvature tensor of the type given by Eq. 12.16 may be defined for it. No attempt is made to treat the stress-strain relations associated with disclinations since this has been extensively done elsewhere (deWit, 1973; Likachev and Kaerov, 1975), at least in the linear approximation. On the other hand, to the extent that we consider disclinations as merely special arrays of dislocations, we can simply employ dislocation theory for such studies. Another consideration which has not been given sufficient attention relates to the presence of disclinations in finite bodies. As in the case of a single dislocation shown in Figure 5.11*b*, it is necessary to distribute an array of surface dislocations over the surface of the body. In the case of disclinations, however the effects are greatly magnified due to the large number of dislocations that in general comprise a disclination. It is in fact the very high stress fields, and their energies, that preclude the presence of dislocations in crystalline materials. This however might not be the case in sufficiently small bodies.

Another basic type of disclination is the twist-type disclination shown in Figure 12.6 (Marcinkowski and Sadananda, 1977). In particular, it is formed by allowing a plastic rotation to occur between two adjacent planes within a circular area. These two planes are denoted by full and open circles. The circular transition zone between rotated and unrotated material shown dashed is termed a twist disclination loop. The twist disclination may be viewed as the disclination counterpart of the screw dislocation, whereas the wedge disclination may be considered as the disclination counterpart of the edge-type dislocation. In Figure 12.6 the strength of the disclination has been shown as $\theta = 53.1°$. Designating this disclination as state (k^{D3}), it can be generated from the reference (K) state by means of the following distortion tensor.

$$A_3^3 = 1 \qquad (12.32a)$$

$$A_1^1 = A_2^2 = \left\{\cos\left(\frac{\theta}{2}\right)H(-x^3)H(r-r_o)\right\}_1 + \left\{\cos\left(\frac{\theta}{2}\right)H(+x^3)H(r-r_o)\right\}_2 \qquad (12.32b)$$

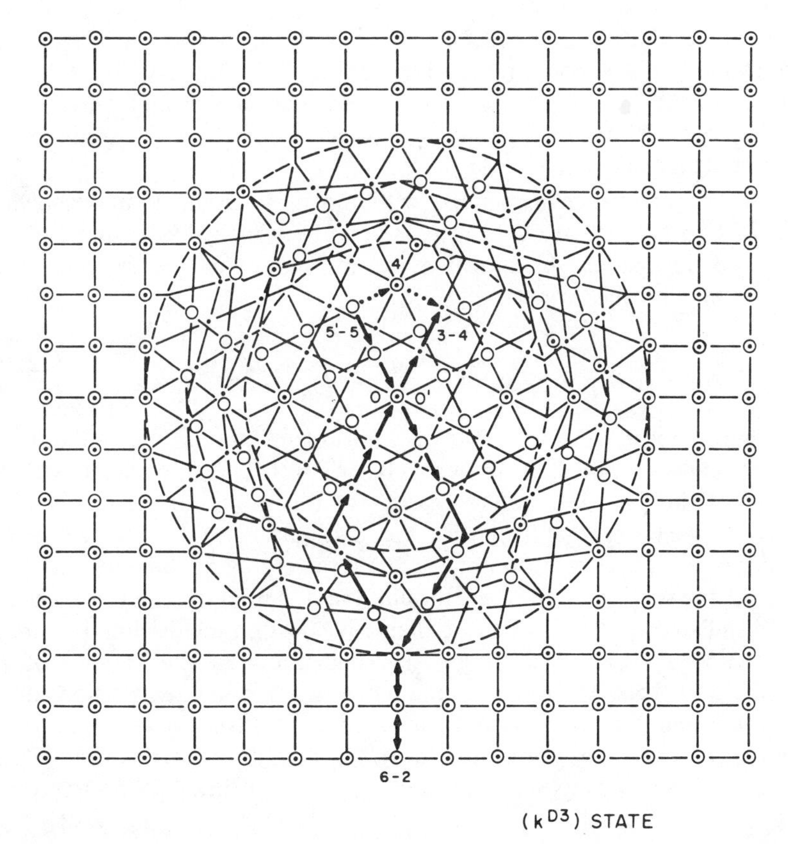

Figure 12.6 Circular twist disclination loop of strength $2\theta = 53.1°$. Reprinted by permission of Springer-Verlag from M. J. Marcinkowski and K. Sadananda, (1977), *Acta Mechanica*, **28**, 159, Figure 1b.

$$A_2^1 = \left\{ \sin\left(\frac{\theta}{2}\right) H(-x^3) H(x^2 - x_o) \right\}_1 + \left\{ -\sin\left(\frac{\theta}{2}\right) H(+x^3) H(x^2 - x_o) \right\}_2 \tag{12.32c}$$

$$A_2^1 = \left\{ -\sin\left(\frac{\theta}{2}\right) H(-x^3) H(x^1 - x_o) \right\}_1 + \left\{ \sin\left(\frac{\theta}{2}\right) H(+x^3) H(x^1 - x_o) \right\}_2 \tag{12.32d}$$

where x_o is the radius of the disclination loop, while the Heaviside function is defined as follows:

$$H(x^2 - x_o) = \begin{cases} 0 \text{ if } (x^2 - x_o) > 0 \\ 1 \text{ if } (x^2 - x_o) < 0 \end{cases} \tag{12.33}$$

with a similar result for $H(x^1 - x_o)$.

It is clear that the imperfect 53.1° twist disclination loop of Figure 12.6 encloses a twist-type grain boundary. The grain boundary may be forced into coherency, in which case it resembles that shown in Figure 10.9, or else it can be completely incoherent in which case it resembles the configuration shown in Figure 10.7*b*.

The Burgers circuit associated with the twist disclination is shown in Figure 12.6, and for greater clarity is again reproduced in Figure 12.7, but in three-dimensions. Analytically, we can write for the Burgers vector associated with this circuit

$$b^1 = -\int_5^0 A_2^1 dx^2 - \int_{0'}^3 A_2^1 dx^2 \tag{12.34a}$$

It is here important to note that the limits of integration are 5–0 and 0′–3 rather than 5–6 and 2–3, respectively, where 0 and 0′ coincide with the center of the disclination. The reason for this is due to the fact that the screw dislocations that comprise the disclination in Figure 12.6 are in reality closed loops that do not include 0 or 0′, that is, the loops lie to the sides of the disclination center. In fact, the nongrain boundary portions of the dislocation loops concentrate themselves within the disclination line or core and thus give rise to the severe bending in the vicinity of the disclination. Thus, in taking the Burgers circuit along those portions of the path 0–6 and 2–0′, just as many negative as positive dislocation segments are crossed, and thus contribute no net dislocation content to the circuit. This, however, is not true along those portions of the path 5–0 and 0′–3

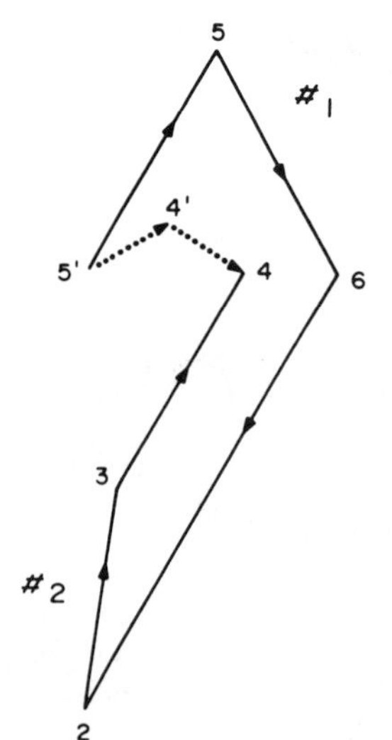

Figure 12.7 Three dimensional portrayal of the Burgers circuit shown in Figure 12.6. Reprinted by permission of Springer-Verlag from M. J. Marcinkowski and K. Sadananda, (1977), *Acta Mechanica*, **28**, 159, Figure 3d.

that measure the net dislocation content comprising the grain boundary. In view of these considerations Eq. 12.32c, when used in conjunction with Eq. 12.34a yields

$$b^1 = \left\{2\sin\left(\frac{\theta}{2}\right)\right\}_1 + \left\{2\sin\left(\frac{\theta}{2}\right)\right\}_2 \equiv \underset{5'-4'}{\Delta x^1} + \underset{4'-4}{\Delta x^1} \qquad (12.34b)$$

as shown by the dotted line segments in Figure 12.6. As discussed in Chapter 6, the factor $\sin\left(\frac{\theta}{2}\right)$ may also be viewed as the product $\tan(\theta/2)$ $\cos(\theta/2)$. An expression similar to that given by Eq. 12.34 can also be used to determine b^2, however, a Burgers circuit at right angles to that shown in Figure 12.6 would have to be utilized.

Rather than use the line integral of Eq. 12.34, we could also have written

$$b^{k^{D3}} = -\int_S S_{l^{D3}m^{D3}}{}^{k^{D3}}\, dF^{l^{D3}m^{D3}} \qquad (12.35)$$

where in view of Eqs. 10.2 and 12.32c

$$S_{23}^{\cdot\cdot 1} = -\frac{1}{2}\bar{A}_2^2\bar{A}_3^3\partial_3 A_2^1 = \left\{\left(\frac{1}{2}\right)\sin\left(\frac{\theta}{2}\right)\delta(x^3)H(x^2-x_o)\right\}_1$$
$$+\left\{\left(\frac{1}{2}\right)\sin\left(\frac{\theta}{2}\right)\delta(x^3)H(x^2-x_o)\right\}_2 \qquad (12.36)$$

which when substituted into Eq. 12.35 yields the same result as that given by Eq. 12.34. Analogous to Eq. 12.7, the dislocation density associated with a twist disclination can also be determined. In addition, from Eq. 12.8, the only nonvanishing component of the Riemann–Christoffel curvature tensor is found to be

$$R_{232}^{\cdot\cdot\cdot 1} = -R_{322}^{\cdot\cdot\cdot 1} = -2\partial_2 S_{12}^{\cdot\cdot 1} = \left\{-\sin\left(\frac{\theta}{2}\right)\delta(x^3)\delta(x^2-x_o)\right\}_1$$
$$+\left\{-\sin\left(\frac{\theta}{2}\right)\delta(x^3)\delta(x^2-x_o)\right\}_2 \qquad (12.37)$$

which, similar to Eq. 12.12 for the wedge disclination, vanishes everywhere except at the disclination line.

The elastic distortion associated with the twist disclination in Figure 12.6 may be removed analogous to that shown in Figure 12.1*c* by tearing along the plane of the disclination. The resulting configuration is shown in Figure 12.8 and is designated as state (a^{D3}). It is important to note that the tearing has allowed the screw dislocations that comprise the disclination loop to run out of the body leaving only compensated screw dislocations in

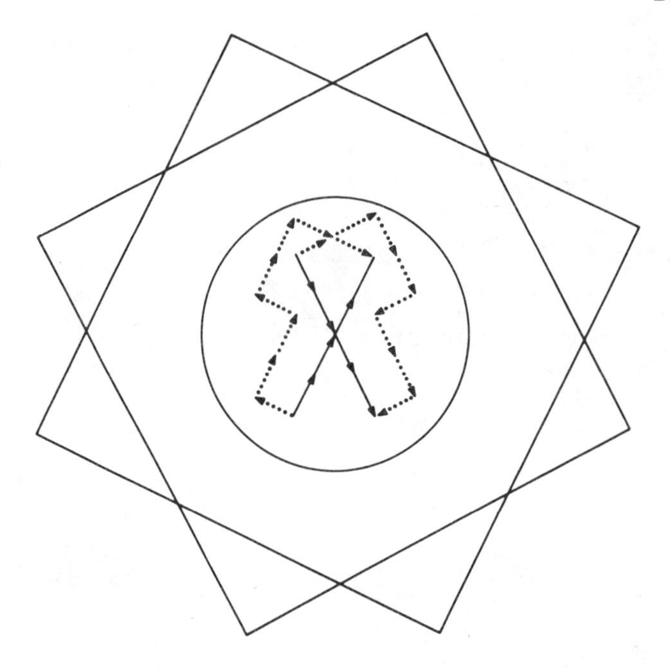

Figure 12.8 Torn counterpart of the twist disclination shown in Figure 12.6.

their wake. This is apparent from Figure 12.8 that shows that each Burgers vector associated with a given screw dislocation is just compensated by another dislocation of opposite sign corresponding to the anholonomic object. Thus, analogous to Eq. 12.20, we may write

$$b^{a^{D3}} = -\int_S \left(S_{c^{D3} b^{D3}}^{\cdot\cdot\ a^{D3}} - \Omega_{c^{D3} b^{D3}}^{\cdot\cdot\ a^{D3}} \right) dF^{c^{D3} b^{D3}} \tag{12.38}$$

where

$$\Omega_{23}^{\cdot\cdot 1} = S_{23}^{\cdot\cdot 1} \tag{12.39}$$

The analysis of a twist disclination can also be linearized in a manner similar to that for a wedge disclination. For a twist disclination, however, the total plastic distortion is the same as a plastic rotation, so that Eq. 12.27 may be rewritten as

$$\beta_{kl}^{P} = \beta_{kl}^{P\omega} \tag{12.40}$$

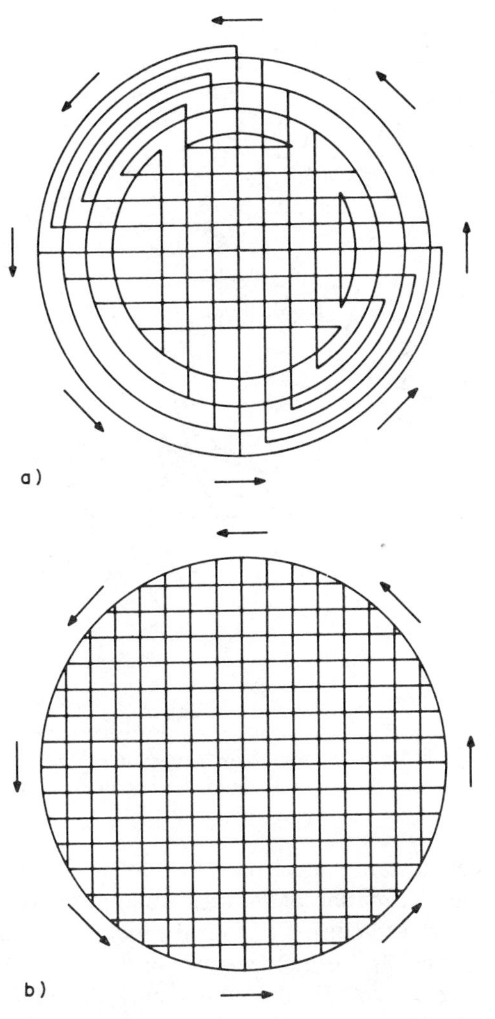

Figure 12.9 Twist disclination loop described in terms of grain boundary dislocations *a*) prior to and *b*) after combination of curved portions of dislocation loops. Reprinted by permission of Springer-Verlag from M. J. Marcinkowski and K. Sadananda, (1977), *Acta Mechanica*, **28**, 159, Figure 5a and b.

We have already seen in Chapters 6 and 10 that when the adjacent grains comprising the grain boundary are forced into a common coincidence, then the Burgers vectors associated with the individual crystal lattice dislocations that comprise the grain boundary can be added to give a grain boundary dislocation. In particular, there is one grain boundary dislocation (GBD) of strength $b_{GB}=2b\cos\theta/2$ spaced every coincidence

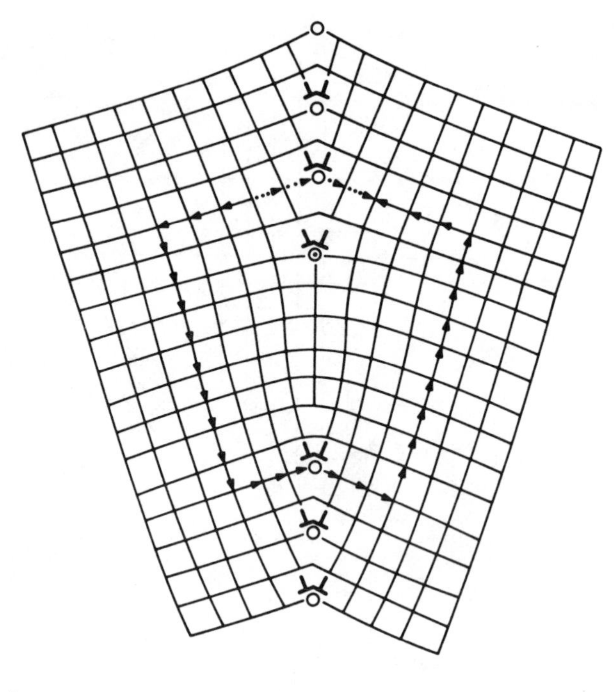

(a)

Figure 12.10a Dislocation deficient tilt boundary of misorientation $2\theta = 53.1°$ represented in terms of the original crystal lattice. Reprinted by permission of Associated Book Publishers Ltd. from M. J. Marcinkowski, *Journal of Materials Science*, in publication, Figure 13a.

site lattice unit cell distance a_{oc} from each other. When represented in terms of GBD's the disclination of Figure 12.6 appears as shown in Figure 12.9*a*. The array is seen to consist of four distinct sets of dislocation loops that form an orthogonal grid near the center of the array. When the curved portions of the loops are forced into coincidence the configuration shown in Figure 12.9*b* obtains. Specifically, the outermost circle comprises the disclination loop to which is attached the orthogonal grid of screw-type GBD's. The twist disclination loop may be viewed as possessing a Burgers vector that is everywhere tangential to the loop as indicated by the arrows in Figure 12.9.

Although single disclinations are of relatively high energy, certain combinations of disclinations such as dipoles possess considerably less energy and are thus more likely to be found in actual crystals (Marcinkowski and Sadananda, 1973). In particular, Figure 12.10*a* shows a grain boundary of misorientation $2\theta = 53.1°$ in which some of the crystal lattice dislocations comprising the grain boundary have been removed. The net result is the

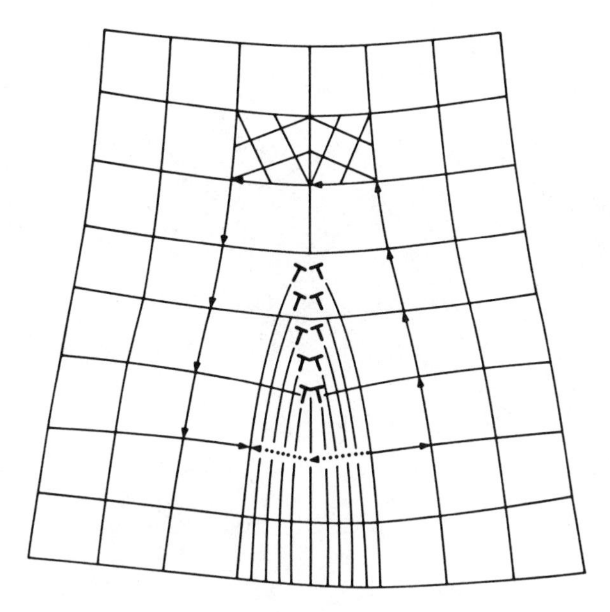

Figure 12.10b Same boundary as that shown in Figure 12.10*a* but represented in terms of the coincidence site sublattice. Reprinted by permission of Associated Book Publishers Ltd. from M. J. Marcinkowski, *Journal of Materials Science*, in publication, Figure 13b.

formation of a wedge disclination dipole, each disclination of which acts as the terminous of a grain boundary. The disclination dipole of Figure 12.10*a* can be redrawn in terms of the coincidence site lattice as shown in Figure 12.10*b*. It is seen that the region between the disclination pair, that originally appeared to be elastically distorted, now appears as an array of dislocations, that is extra half planes that are represented in terms of the coincidence site lattice sublattice. This is in keeping with the findings in Chapter 7 whereby it was shown that an elastic distortion may always be transformed into a dislocated state by employing a trasformation which gives rise to a common metric. In particular, the coincidence site lattice provides the common metric. Analogous to the wedge disclination dipole of Figure 12.10*a*, a twist disclination dipole can also be constructed as is shown in Figure 12.11*a* in terms of the original crystal lattice. In the present case, the twist disclinations are drawn as a pair of dashed straight lines. Again they enclose elastically distorted material. When redrawn in terms of the coincidence site lattice, as shown in Figure 12.11*b*, the elastically strained region between the disclination pair comprising the dipole is seen to consist of an array of horizontal screw-type dislocations.

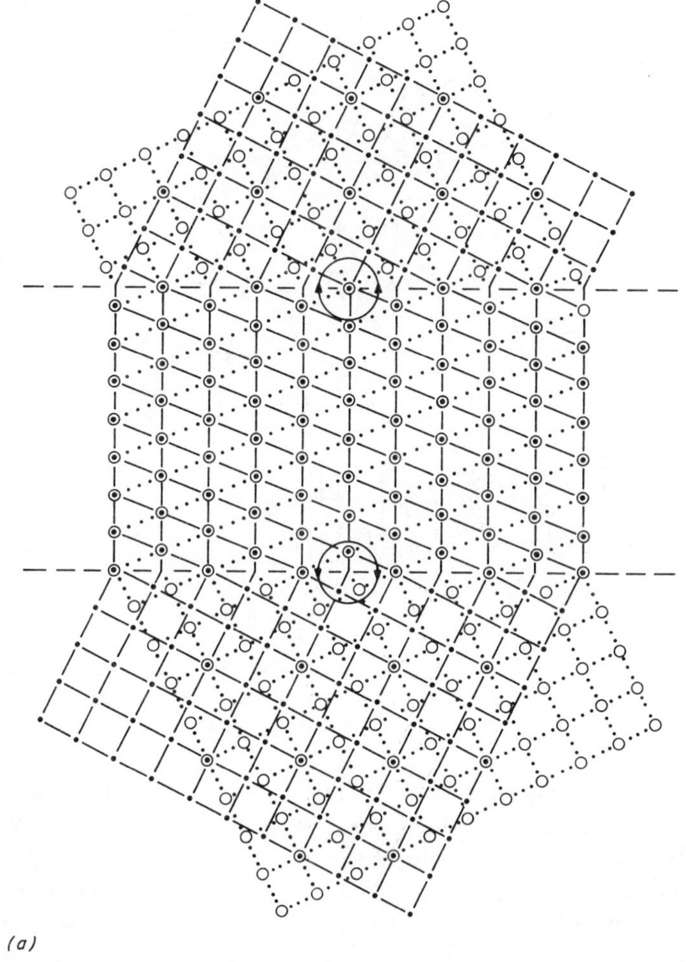

Figure 12.11a Twist disclination dipole. Reprinted by permission of Springer-Verlag from M. J. Marcinkowski and K. Sadananda, (1977), *Acta Mechanica*, **28**, 159, Figure 6a.

REVIEW

The two basic types of disclinations have been treated in this chapter, that is, the wedge and the twist type. In the former case, the disclination may be viewed as a tilt-type grain boundary which terminates abruptly within the interior of the body such as is shown in Figure 12.1*a*. This may be termed a wedge disclination of strength 53.1° since it is apparent from this figure that the disclination may be visualized as being formed by the

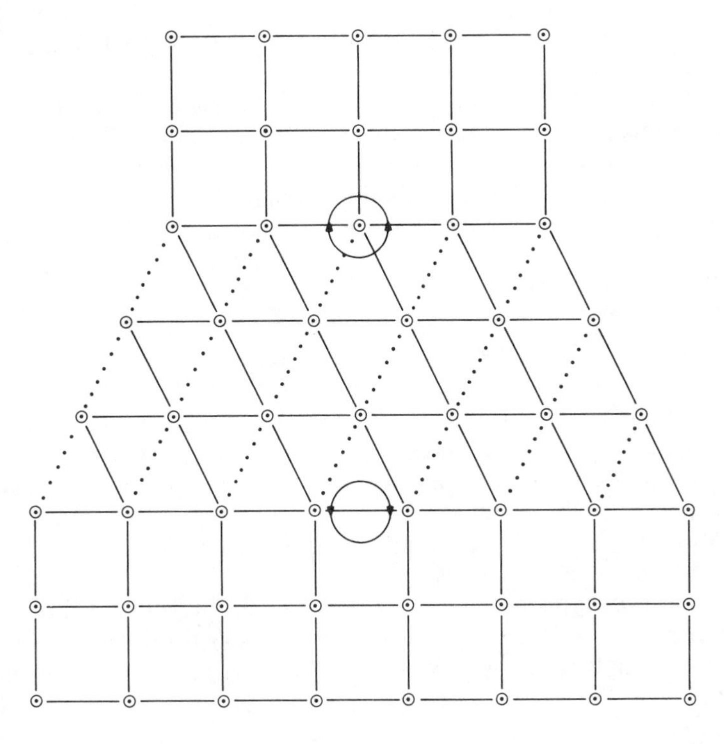

Figure 12.11b Coincidence site lattice representation of the twist disclination dipole shown in Figure 12.11*a*. Reprinted by permission of Springer-Verlag from M. J. Marcinkowski and K. Sadananda, (1977), *Acta Mechanica*, **28**, 159, Figure 6b.

introduction of a wedge 0–2′–2″–3′–0 of angle 53.1° into an otherwise perfect crystal. The wedge disclination in Figure 12.1*a* may also be termed a partial or imperfect disclination since the angle 53.1° does not correspond to a symmetry operation within the crystal. On the other hand, Figure 12.3*b* shows a perfect 90° wedge disclination. Both the perfect and imperfect disclinations may be viewed as consisting of dislocations. However, whereas in the case of imperfect disclination the dislocations are discretely distributed, those associated with a perfect disclination are continuously distributed. The consequences arising from this difference are that both the localized stress fields as well as the free surfaces about the dislocations comprising the partial disclination are finite, whereas they vanish in the case of the perfect disclination. It is perhaps these latter considerations that have led some to view a perfect disclination as not terminating on a grain boundary.

As Eq. 12.1 shows, the distortion tensors associated with disclinations are very similar to those associated with with grain boundaries, as comparison with Eq. 6.1. However, in the former case, the distortions are in the form of products of Heaviside functions, where the extra step function is associated with the termination of the grain boundary within the crystal. These distortions allow a well-defined torsion tensor to be associated with a disclination as can be seen by reference to Eq. 12.6. A corresponding dislocation density can in turn be obtained from this torsion tensor as shown by Eq. 12.7. Another interesting aspect associated with a disclination is that it also possesses a curvature tensor as given by Eq. 12.12. The physical difference between the torsion tensor and the curvature tensor lies in fact that in the former case a closure failure is the result of the displacement of a point about a given circuit, whereas in the latter case the closure failure is the result of the displacement of a vector about that circuit. More specifically, the vector does not coincide with itself at the initial and final points in the circuit.

A partial twist-type disclination of strength 53.1° is illustrated in Figure 12.6. It may be viewed as a circular loop that acts as the terminus of a twist type grain boundary. Similar to that of a tilt boundary, the various tensor quantities associated with twist-type disclinations can be readily obtained. In addition, various combinations of disclinations, such as dipoles, can be formed which possess very much lower energies than do single disclinations. It is thus shown that disclinations provide a very powerful concept which enables a much deeper generalization and thus understanding of defects, in general.

CHAPTER 13

Applications to Specific Problems

Nearly all of the material covered thus far has been devoted toward laying a solid theoretical foundation with which to solve nearly all of the important problems associated with the mechanical behavior of matter. In order to demonstrate the power of these methods, three of the most important problems associated with materials are considered. First, the exact nature of cracks, both elastic and plastic, is analyzed. Second, the exact structure of a grain boundary of arbitrary misfit is treated. Finally, the precise formulation of the concept of surface tension in terms of surface dislocations in both solids and liquids is presented.

The previous chapters have shown quite clearly how any given state or distortion, either elastic or plastic can be described in terms of some suitable distribution of dislocations. In principle, these analyses can be carried out analytically since the concepts are exact and contain no assumptions or approximations. In actual practice, on the other hand, the mathematical methods become so cumbersome or difficult so as to render such approaches undesirable. Fortunately, however, the methods of computer analysis allow such calculations to be carried out in a rather straightforward manner (Marcinkowski, 1971). In order to demonstrate the power of the present techniques we apply them to the study of three very important classes of problems; the analysis of the fracture process, the structure and energy of grain boundaries and two-phase interfaces, and the nature of surface tension in liquids and solids.

13.1 THE FRACTURE PROCESS

We are now in a position to treat what is perhaps the economically most important mechanical property of all, namely that of fracture. As we shall see, no new concepts are required, and it is only necessary to apply those concepts already developed. In particular, consider the limiting case of the stressed hole in Figure 5.15*b* as its vertical dimension approaches zero. The resulting configuration is shown in Figure 13.1*a* and is seen to be simply a tensile of Mode I crack. Again, in order to insure that the surface tractions vanish on the faces of the crack, it is necessary to continuously distribute surface dislocation dipoles over those faces (Marcinkowski and Das, 1974). As indicated in Figure 13.1*a*, that portion of the dipole shown dotted represents a stress-free dislocation that is associated with the vertical surface segment, while that portion of the dipole drawn solid has associated with it a stress field. The vacant space in Figure 13.1*a* can next be filled with stress-free extra matter, similar to that described in connection with Figure 5.19*d* to generate the configuration shown in Figure 13.1*b*. This operation eliminates all of the newly created free surfaces associated with the tensile crack in Figure 13.1*a*. If now the externally applied stress σ_{yy} is removed the configuration shown in Figure 13.1*c* obtains. In particular, we are simply left with a special array, that is, pile-up of prismatic type dislocations. In the creation of the surface dislocations in Figure 13.1*a* the surface energy or tension is taken to be zero. If, on the other hand, this surface energy is allowed to become arbitrarily large, then the surface dislocation dipoles collapse to give the configuration shown in Figure 13.1*d*. In this case the stress does not vanish on the surface but is borne by the surface in the form of a surface tension.

A more complete description of the Mode I crack is shown in Figure 13.2*a* where the primary type surface dislocations are drawn solid while the secondary array is shown dotted (Jagannadham and Marcinkowski, 1978e). The secondary array is seen to be similar to that associated with the top and bottom faces of the hole in Figure 5.18. A shear or Mode II type crack is shown in Figure 13.2*b* where again both primary and secondary edge type surface dislocations are needed to satisfy the stress-free boundary conditions. These arrays are similar to those associated with the sheared hole shown in Figures 5.23 and 5.25*b*. Finally, Figure 13.2*c* shows a Mode III type crack and is seen to consist entirely of a primary array of screw-type surface dislocations. Furthermore, unlike that of the Mode I and II cracks, the faces of the Mode III crack are planar and there is no secondary dislocation array.

Closely related to the crack configurations in Figure 13.2 are the dislocation arrays shown to the right of each of these configurations and denoted by primed letters. Unlike the crack dislocations, that are in effect surface dislocations, the lattice dislocations can exist in the material even after the external load is removed. The reason for this is that the dislocations in Fig. 13.2*a′* can only annihilate with one another if climb is allowed

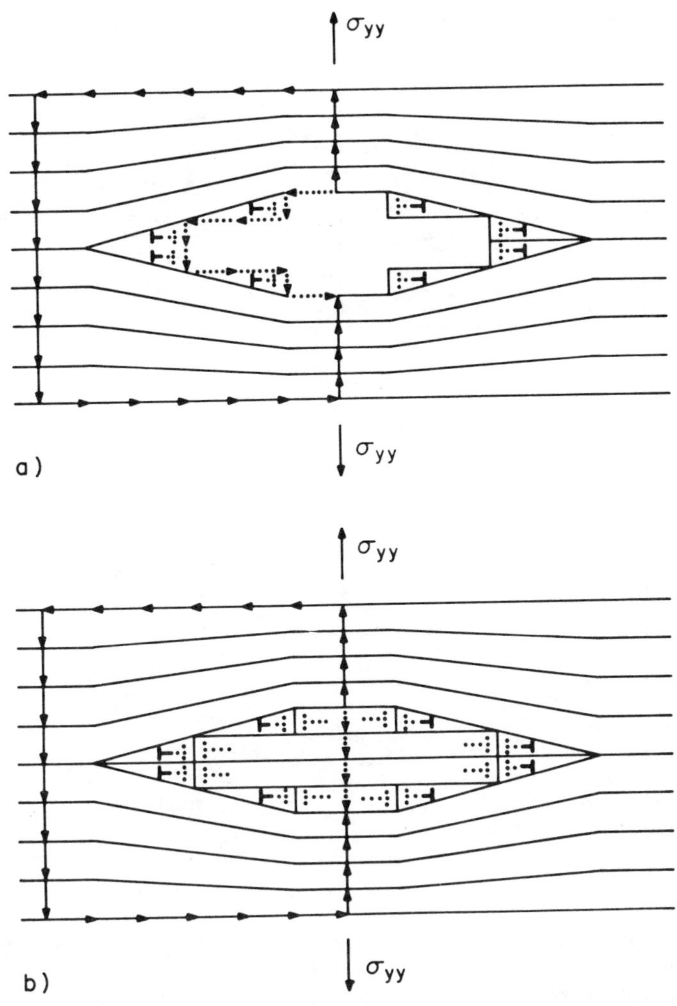

Figure 13.1 *a*) Tensile or Mode I crack in a body subjected to an applied stress σ_{yy}. Reprinted by permission of Springer-Verlag from M. J. Marcinkowski, (1978e), *Acta Mechanica*, in publication, Figure 5c.

to occur to which there is a resistance, especially at low temperatures. In the case of Figures 13.2*b*′ and 13.2*c*′, it is the lattice friction stress which opposes the glide motion of these dislocations and thus prevents mutual annihilation. The dislocation arrays shown towards the right in Figure 13.2 are also important in that upon application of an external load, they provide the nucleus for a crack. This is easy to see in the case of Figure

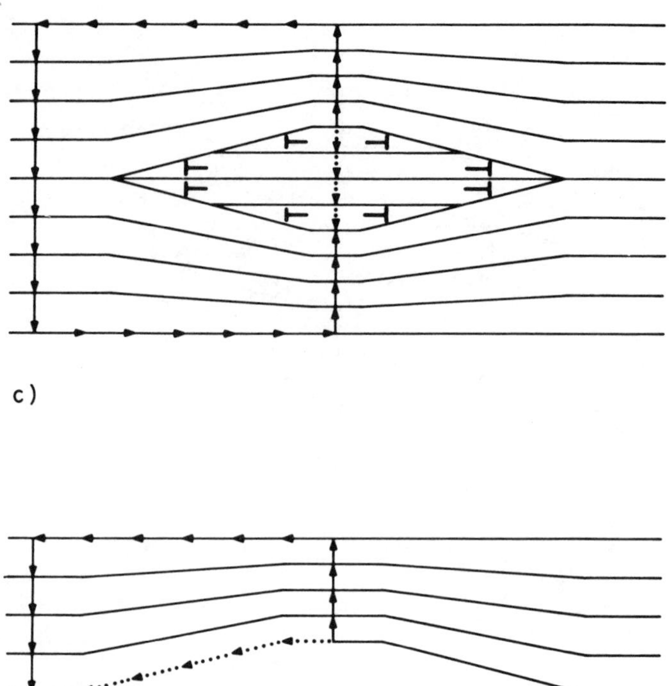

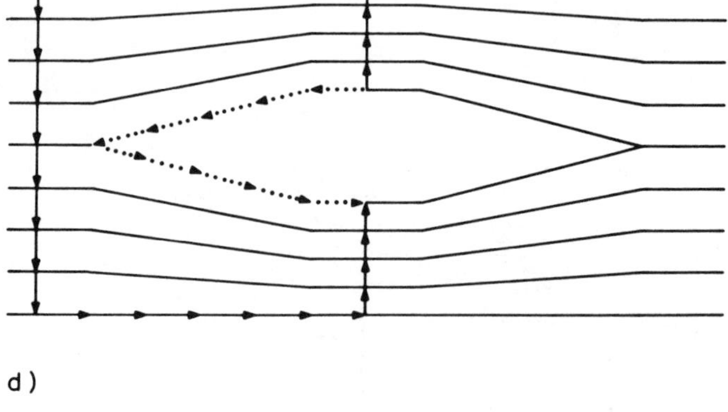

Figure 13.1 (Continued) *c*) Same as *b*) but after removal of the applied stress σ_{yy}. *d*) Same as *a*), but after elimination of surface dislocations. Reprinted by permission of Springer-Verlag from M. J. Marcinkowski, (1978e), *Acta Mechanica*, in publication, Figure 5b.

13.2*a* which upon application of a load attains a configuration shown in Figure 13.1*b* that can grow into a crack with no additional expenditure of energy. More is said with respect to this important point shortly. It is also apparent from Figure 13.1 that a completely self-consistent Burgers circuit can be associated with a given crack configuration. In particular, the vertical dotted closure failures in Figure 13.1*a* correspond to the Burgers vectors of the surface dislocations and thus have associated with them a torsion tensor. The horizontal dotted closure failures, on the other hand,

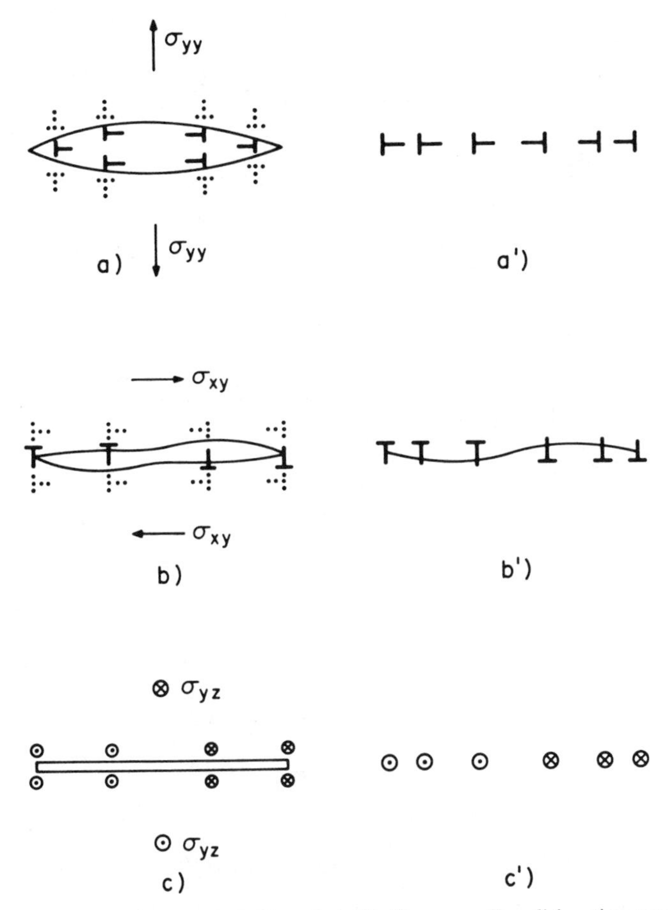

Figure 13.2 Cracks of type *a*) I, *b*) II, and *c*) III. Corresponding dislocation arrays of type *a'*) I, *b'*) II, and *c'*) III.

correspond to newly created free surface and thus possess only an anholonomic object. In the case of Figure 13.1*b*, since all of the free surfaces are eliminated by the introduction of extra matter, the closure failures correspond to the presence of dislocations, that is, torsion. The same also holds in the case of Figure 13.1*c*. For Figure 13.1*d*, since the high surface energy has eliminated the surface dislocations, the dotted closure failures correspond only to the creation of free surfaces, that is, anholonomic object.

We are now in a position to consider the energy of the crack arrays shown in Figure 13.2. For simplicity, we consider the shear crack shown in Figure 13.2*b* in the linear approximation. Under these conditions, the

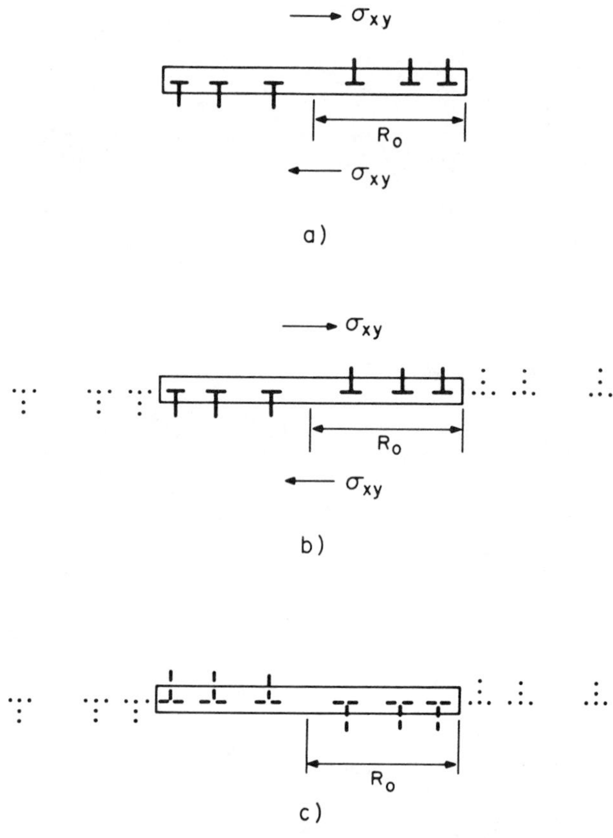

Figure 13.3 Loaded *a*) elastic and *b*) plastic shear crack. *c*) Unloaded plastic shear crack.

primary dislocation array can be assumed to be planar and the secondary array omitted since it has been shown that this array contributes only a small part of the total energy of the crack (Marcinkowski and Das, 1974). The shear crack is reproduced again in Figure 13.3*a* and may be termed an elastic shear crack since there is no plasticity, that is, crystal lattice dislocations associated with this crack. The total energy of an crack of radius R_o can be written as

$$E_T = E_S + E_I - E_\sigma + E_\gamma \tag{13.1}$$

where E_S is the self energy of all of the crack dislocations comprising the crack, E_I the interaction energy between all of these crack dislocation, while E_σ is the work associated with the applied stress, σ_{xy}. It may be written as

$$E_\sigma = 2 \sum_{i=1}^{N} \sigma_{xy} b R_i \tag{13.2}$$

where b is the Burgers vector of each crack dislocation, R_i the position of each dislocation, while N is their total number. The last term in Eq. 13.1 corresponds to the surface energy of the crack which may be written as

$$E_\gamma = 2(2R_o)\gamma \tag{13.3}$$

where γ is the free surface energy per unit area. In spite of the fact that the crack dislocations in Figure 13.3*a* correspond to a continuous distribution of infinitesimal dislocations, we nevertheless treat them as discrete. This makes the problem more amenable to computer techniques which can be carried out to any degree of accuracy desired by subdividing the Burgers vector of the crack dislocation to ever smaller values. In order to solve Eq. 13.1 it is first necessary to determine the equilibrium positions of the crack dislocations for a given value of R_o. This may be done by minimizing the energy E given by

$$E = E_S + E_I - E_\sigma \tag{13.4}$$

as a function of position of the crack dislocations. Crack dislocations are continuously added to the crack so long as E given by Eq. 13.4 decreases. This operation is equivalent to ensuring that the surface tractions vanish over the crack faces. The computer techniques employed in minimizing the energies of various dislocation arrays are well documented (Marcinkowski, 1971). Once the crack positions are determined, E_T given by Eq. 13.1 can be evaluated. The resulting relation between E_T and R_o is shown by the lowermost curve in Figure 13.4 where the maximum in the curve corresponds to that of the Griffith condition. In particular, cracks with size greater than the Griffith size grow under constant applied stress. Now in general, the energies to create a crack in a perfect crystal are too high to be provided by processes such as thermal activation. It is therefore reasonable to assume that preexisting free surfaces, that is, cavities equal to or greater than the Griffith size are present or else that certain combinations of dislocations that resemble a crack dislocation array, such as shown in Figure 13.1*b*, are present within a real solid. In this latter case, the configuration could also grow spontaneously as a Griffith crack if present above a certain critical Griffith size.

It is now possible to develop an important relationship between the elastic energy released by the formation of an elastic crack and its

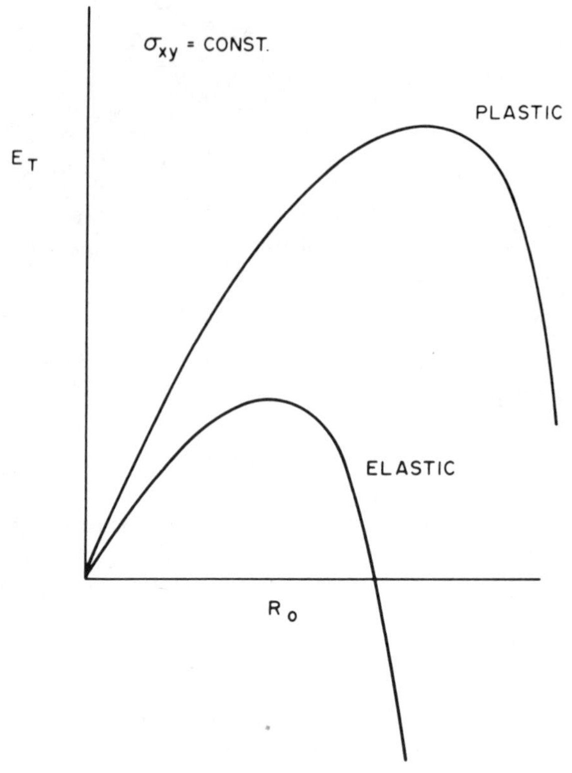

Figure 13.4 Energy versus crack size for an elastic and a plastic crack.

dislocation content. In particular, we can write Eq. 13.1 alternately as

$$E_T = E_\gamma - E_e \tag{13.5}$$

where E_e is the elastic energy released by the crack. Combining Eqs. 13.5 and 13.1 yields

$$E_S + E_I - E_\sigma = -E_e \tag{13.6}$$

The elastic strain energy released by the crack can also be written as

$$E_e = \left(\tfrac{1}{2}\right) 2 \sum_{i=1}^{N} \sigma_{xy} b R_i \tag{13.7}$$

which when combined with Eq. 13.2 yields

$$E_\sigma = 2E_e \tag{13.8}$$

When the last equation is in turn combined with Eq. 13.6, we obtain

$$E_e = E_S + E_I \tag{13.9}$$

We thus have the important result that the elastic energy released by the crack is equal to the self and interaction energies of the crack dislocations.

Let us now consider the plastic shear crack which is shown in Figure 13.3*b*. In this case, the crack dislocations are allowed to move out of the crack tip into the crystal lattice to become crystal lattice dislocations (Marcinkowski and Das, 1971). In so doing they move against a lattice friction stress or resistance σ_f. Under these conditions Eq. 13.1 must be modified as follows

$$E_T^P = E_S + E_I - E_\sigma + E_\gamma + E_f \tag{13.10}$$

where E_f is the work done against the lattice friction given by

$$E_f = 2\sum_{i=1}^{M} \sigma_f b(R_i - R_o) \tag{13.11}$$

where M is the number of crystal lattice dislocations that comprise the plastic zone of the crack tip. Again, before Eq. 13.10 can be evaluated, it is first necessary to determine the positions of all the crack and crystal lattice dislocations in Figure 13.3*b* for a given R_o. This is done by minimizing the following energy as a function of these dislocation coordinates

$$E^P = E_S + E_I - E_\sigma + E_f \tag{13.12}$$

When this is done, the uppermost curve shown in Figure 13.4 obtains. We see at once that the energy of a plastic crack is much greater than that of the corresponding elastic crack in agreement with what was generally to be expected (Sadananda, Jagannadham and Marcinkowski, 1978). It is important to note that in obtaining the E_T versus R_o curve for the plastic tensile crack, it was necessary to cumulatively add all of the friction energy contributions to the curve from $R_o = 0$ to its given size, that is, continuous plastic crack. One might think that an alternate procedure would be to allow the crack to grow elastically to R_o followed by the formation of a plastic zone, that is, discontinuous plastic crack (Marcinkowski and Das, 1971; Marcinkowski, 1975). The former case is more realistic since it allows the crack to release crystal lattice dislocations continuously during its growth process. We are again confronted with the problem of how the large activation energies for plastic crack formation in an otherwise perfect crystal are attained. Again, we assume the existence of preexisting cavities or dislocation arrays of critical size. One can imagine this as being

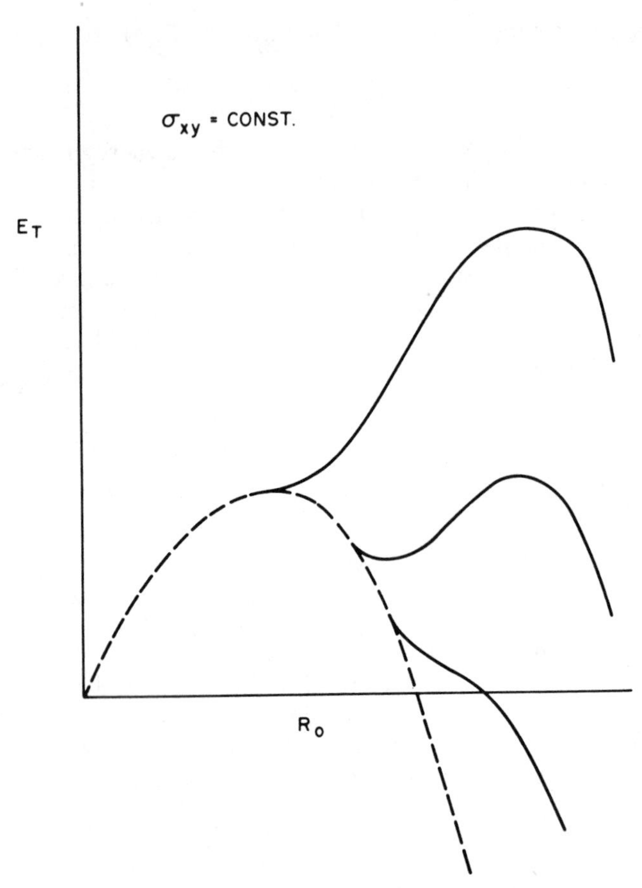

Figure 13.5 Energy versus crack size for a plastic crack (solid lines) that forms from an elastic crack (dashed line).

equivalent to already lying somewhere on the elastic energy curve shown dashed in Figure 13.5. If the preexisting crack corresponds to the maximum in the dotted E_T versus R_o curve, then in order for it to proceed to grow plastically, the energy would continue to rise to another maximum in the manner shown by the topmost solid curve in Figure 13.5. We thus have the important result that an elastic Griffith crack cannot spontaneously continue to grow plastically. If, however, the preexisting crack were sufficiently larger than the Griffith size, then the crack could continue to grow spontaneously in a plastic manner with a continual reduction in energy as shown by the bottommost solid curve of Figure 13.5. There is a way however that an elastic Griffith crack can grow spontaneously as a plastic crack and this is by increasing the externally applied stress such as shown in Figure 13.6. In this figure, $\sigma_3 > \sigma_2 > \sigma_1$, so that at an applied stress of σ_3, the elastic crack can grow plastically with a continuous decrease in energy.

An equation similar to that given by Eq. 13.8 for the elastic energy release associated with a plastic crack can be determined by first writing analogous to Eq. 13.5

$$E_T^P = E_\gamma + E_f - E_e^P \tag{13.13}$$

that when combined with Eq. 13.10 yields an expression identical to that given by Eq. 13.6. The elastic energy released by the plastic crack can also be written, analogous to Eq. 13.8, as

$$E_e^P = \left(\tfrac{1}{2}\right) 2 \sum_{i=1}^{N} (\sigma_{xy} - \sigma_f) b R_i \tag{13.14}$$

that when combined with Eq. 13.2 yields

$$E_\sigma = 2E_e^P + E_f \tag{13.15}$$

Combining Eqs. 13.15 and 13.6 gives

$$E_e^P = E_S + E_I - E_f \tag{13.16}$$

Comparing this result with Eq. 13.9, we see that the frictional energy "robs" elastic energy from the crystal.

The elastic energy released in the formation of an elastic shear crack can be written exactly as

$$E_e = \frac{\pi(1-\nu)}{2\mu} \sigma_{xy}^2 R_o^2 \tag{13.17}$$

Upon combination of Eqs. 13.3, 13.5, and 13.17, and making use of the Griffith condition that $dE_T/dR_o = 0$, we obtain

$$R_{oG} = \frac{4\gamma\mu}{\pi(1-\nu)\sigma_{xy}^2} \tag{13.18}$$

for the Griffith crack size. It has been argued that for a plastic crack, γ should be replaced by $(\gamma + \gamma_p)$ where γ_p is a plastic work term and is usually much greater than γ. This assumption however leads to some

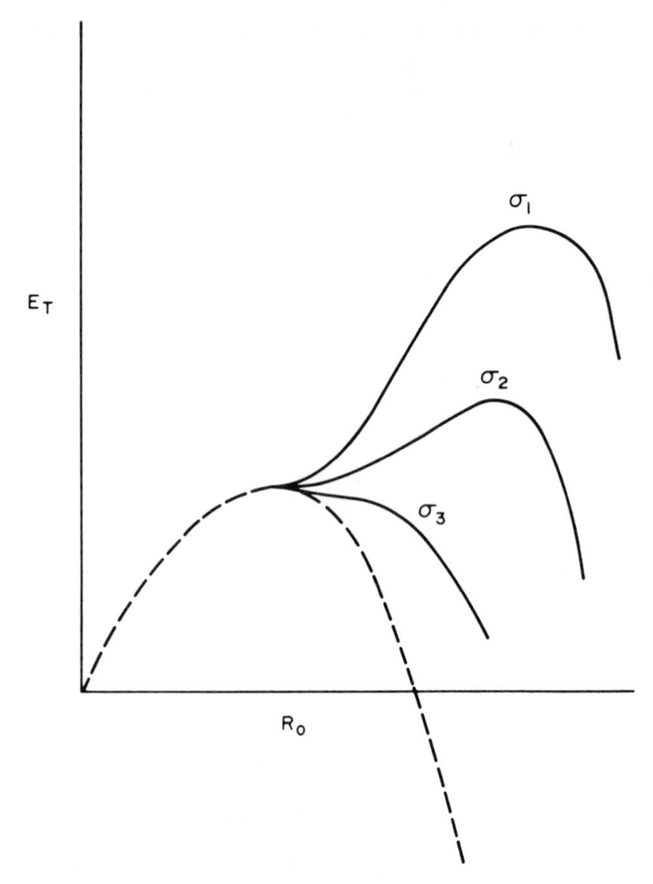

Figure 13.6 Energy versus crack size for a plastic crack (solid lines) that forms from an elastic crack (dashed line) under varying applied stress.

conceptual difficulties. In particular, the crystal lattice dislocations which comprise the plastic zone in Figure 13.3*b* interact strongly with the crack dislocations. It is therefore not possible to uniquely define an energy contribution arising from the plastic zone.

It is next of interest to analyze what happens when a plastic crack is unloaded (Jagannadham and Marcinkowski, 1978f). A resulting configuration is shown in Figure 13.3*c*. Here we see that the crack dislocations have undergone mutual annihilation amongst themselves. Their place in turn is taken by dislocations of opposite sign which are drawn dashed. These may be referred to as anticrack dislocations and arise in order to satisfy the stress-free boundary conditions on the faces of the crack that arise from the crystal lattice dislocations which are prevented from moving back into the crack by the lattice friction stress. Actually, some of the crystal lattice dislocations are pulled back into the crack and the crack shrinks somewhat. In terms of the E_T versus R_o curve of Figure 13.7, the unloading process corresponds to the path 1-2. The energy increases, since upon unloading the plastic crack no longer releases elastic strain, but instead contributes to it. The crack may be subsequently reloaded along the path 2–3 in Figure 13.7 where it reaches a minimum at point 3 with no further growth possible, since the energy begins to increase. Upon subsequent unloading to point 4, followed by loading to point 5, another stable minimum is reached. Finally, upon unloading to point 6 followed by a reloading no further minima are reached and the crack grows without limit. It is obvious that we have just given the precise mechanism for the growth of a subcritical Griffith crack into a propagating crack by means of a fatigue process (Jagannadham and Marcinkowski, 1978g). Physically

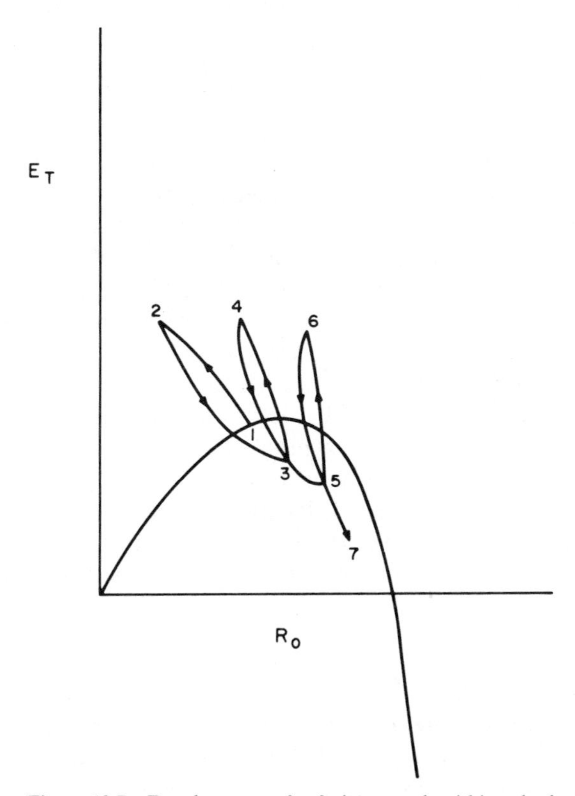

Figure 13.7 Development of a fatigue crack within a body.

what happens is that after each unloading, more crystal lattice dislocations are left in the plastic zone. It thus becomes easier to make the crack grow upon subsequent loading since these crystal lattice dislocations are already present in the lattice and thus no additional frictional energy is required to bring them to this position. Another important feature of the fatigue crack is that it is fully history dependent. In particular, it never returns to its original state upon loading or unloading. This history dependence is entirely the result of the friction stress and is in fact the basis of the fatigue process. An elastic crack, on the other hand, is fully reversible and can be shrunk to zero or increased without end along the same paths, that is, along the E_T versus R_o curve of Figure 13.4. The lattice friction stress and its relation to history dependent processes has been considered in detail in a number of studies (Sadananda and Marcinkowski, 1972; Sadannanda, 1976).

13.2 CALCULATION OF GRAIN BOUNDARY ENERGIES

We are now in a position to calculate the energy of a grain boundary which is quite unique. This gives insight into the nature of such boundaries that has hitherto been lacking. In particular, Figure 13.8 shows a reproduction of the grain boundary of Figure 7.5 in much greater detail. We see that the boundary consists of coalesced segments that have formed in order to reduce the amount of free surface energy. In a sense, the grain boundary may be viewed as consisting of an array of asymmetric tensile-type cracks. In order to satisfy the free-surface boundary conditions on the crack surfaces, a set of surface dislocation dipoles must be places on these surfaces as shown in the centermost crack of Figure 13.8. For clarity, the surface dislocations are indicated by a single symbol for the two outermost cracks. Again, as in Figure 13.1*a*, it is only that portion of the dipole indicated by a solid dislocation symbol which provides a stress field. The dotted dislocations or ledge portions of all the dipoles just balance with the ledge of the grain boundary dislocations which are also indicated by dotted symbols. In both cases, these possess no stress field. Similar to the case of a crack, the total energy of the grain boundary in Figure 13.8 can be written as

$$E_T = E_S + E_I + E_\gamma \tag{13.19}$$

which, except for the absence of the term E_σ, is similar to that given by Eq. 13.1. The energy E_T in Eq. 13.19 may be minimized numerically with respect to the length L in Figure 13.8 for a given value of R. It is then

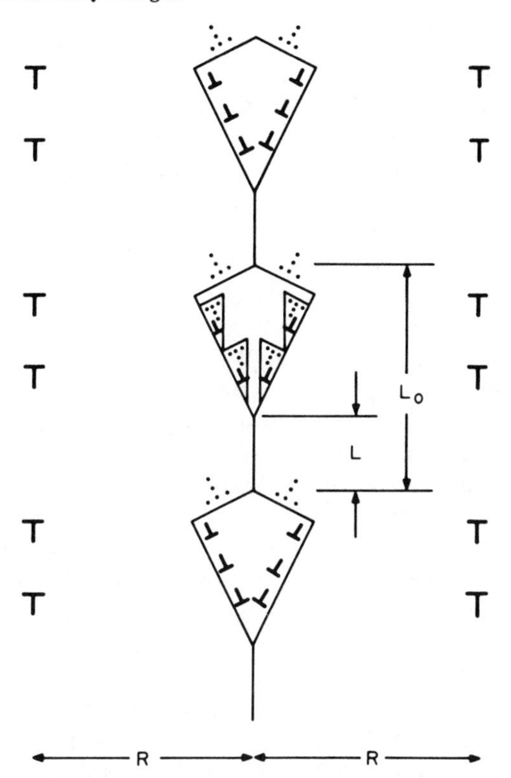

Figure 13.8 Surface dislocation representation of a grain boundary. Reprinted by permission of Internationaler Buch-Versand GmbH from K. Jagannadham and M. J. Marcinkowski (1978h), *Physica Status Solidi*, **a50**, Figure 2.

found that E_T possesses a minimum for some critical value of $R \simeq L_o$ where the assumption is made that the angles defining the shape of the crack are preserved at all times (Jagannadham and Marcinkowski, 1978h). It is apparent that the dislocation arrays shown in Figure 13.8 are in effect dipoles and are actually a somewhat more accurate description of the configuration shown in Figure 7.4*a*. The dipoles are prevented from annihilating with one another by the surface tension of the grain boundary surface energy which increases as R decreases. The short range nature of the stresses about a grain boundary are thus readily explained by this model. Furthermore, because of the short range nature of the stress fields involved, that is, on the order of the separation between cracks, it is only necessary to consider the interaction energies between a given crack and its two nearest neighbors as shown in Figure 13.8. The sum of the Burgers vectors of all of the crack dislocations also remains constant, which means that as L increases, the surface dislocations are packed together more tightly. It is now interesting to note that as the surface energy goes to zero, the surface dislocation dipoles in Figure 13.8 annihilate with one another, in turn allowing L to go to zero so that the completely torn grain boundary configuration shown in Figure 6.1*b* obtains. On the other hand, as the surface energy goes to infinity, the free surface or dotted portions of the dislocations in Figure 13.8 combine with one another with resultant annihilation. This is synonymous with the elimination of all the free surfaces, so that the grain boundary configuration shown in Figure 6.1*a* obtains. Thus, only the strain-free, that is, solidly drawn portions of the surface dislocation dipoles remain. We thus have an accurate or exact continuum representation of the structure of a grain boundary.

The energy of the grain boundary shown in Figure 6.1*b* can be reduced considerably by allowing the grain boundary dislocations to halve their strength in the manner shown in Figure 13.9*a*. In the discrete atomistic picture, this process may be viewed as equivalent to the insertion of an atom into the grain boundary voids of Figure 6.1 to generate the configuration illustrated in Figure 13.9*b*. It is obvious that because of geometric considerations, the filling cannot be as perfect as that shown for the 90° boundary of Figure 7.6*b* so that the grain boundary dislocation array must remain discrete rather than continuous. The calculation for the partially coalesced counterpart of Figure 13.9*a* still remains the same as that associated with Figure 13.8 except that now the total strength of all the surface dislocations within each crack must be halved, while at the same time reducing the distance L_o between adjacent cracks by this same factor. The fully coalesced state of Figure 13.9*a* is shown in Figure 13.9*c*. Here it is somewhat more clear that unlike the case in Figure 6.1*a*, the dislocations in Figure 13.9*c* are not crystal lattice dislocations. This means that they cannot glide back into the crystal since their Burgers vector is in effect only half that of a crystal lattice dislocation. In effect, the dislocations shown in Figure 13.9*c* are the small grain boundary dislocations into which the larger ones shown in Figure 6.1*a* decompose, that is,

$$\boldsymbol{b}_{GB} \rightarrow \tfrac{1}{2}\boldsymbol{b}_{GB} + \tfrac{1}{2}\boldsymbol{b}_{GB} \tag{13.20}$$

where $|\boldsymbol{b}_{GB}| = 2b\cos\theta/2$.

The self energy per unit length of an infinite straight edge-type dislocation is given by

$$E_S = \frac{\mu b^2}{4\pi(1-\nu)}\ln\left(\frac{R}{r_o}\right) \tag{13.21}$$

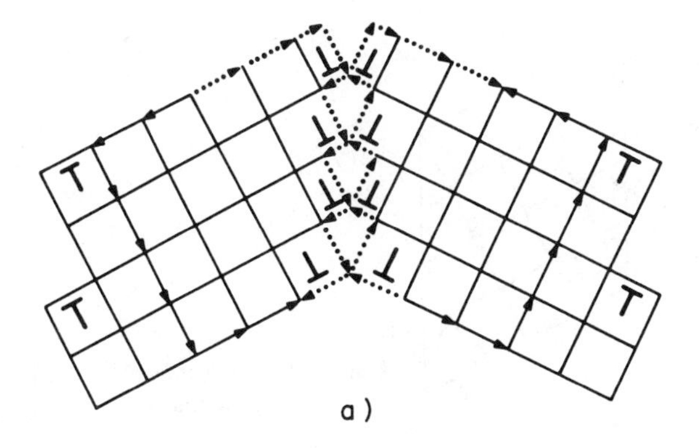

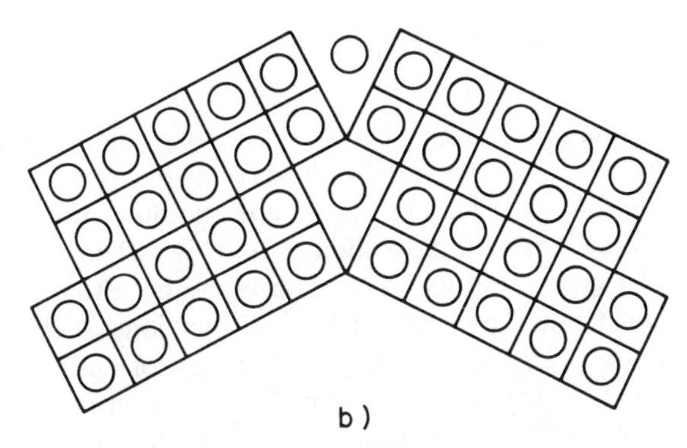

Figure 13.9 *a*) Symmetric tilt boundary similar to that shown in Figure 6.1*b*, but after dissociation of grain boundary dislocation to a lower energy configuration. *b*) Atomistic representation of *a*).

where r_o is the core radius of the dislocation and R is on the order of the crystal dimensions. A similar equation holds for the screw dislocations upon omission of the term $(1-\nu)$. The interaction energy between a pair of edge-type dislocations of opposite sign separated by a distance d, which makes an angle θ with respect to the direction of b is given by

$$E_I = -\frac{\mu b^2}{2\pi(1-\nu)}\left[\ln\frac{R}{d} - \sin^2\theta\right] \qquad (13.22)$$

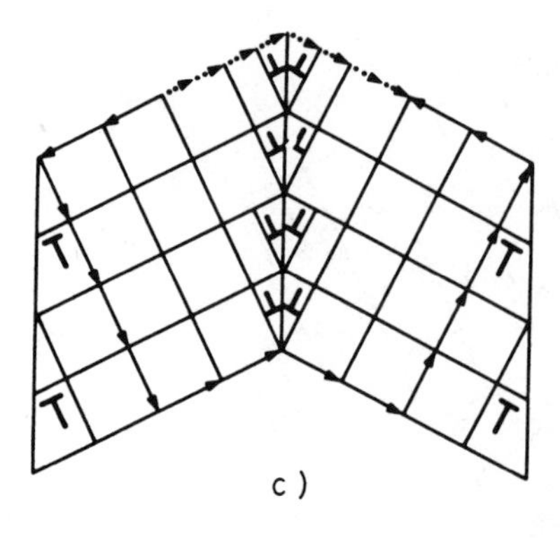

Figure 13.9 (Continued) *c*) Fully coalesced version of the tilt boundary shown in Figure 13.9*a*.

Now in the case of an edge dislocation dipole, the total energy can be written as

$$E_T = 2E_S + E_I = \frac{\mu b^2}{2\pi(1-\nu)}\left[\ln\frac{d}{r_o} + \sin^2\theta\right] \tag{13.23}$$

This is an important result in that it shows that the total energy of a dislocation dipole is independent of the crystal dimensions R. This is a general result and holds whenever there are an equal number of dislocations of opposite sign within a body, that is, when the law of the conservation of Burgers vectors holds. This of course also includes the case of a crystal dislocation and its surface dislocations such as shown in Figures 11.1*a* and 11.2*a*. In writing the energy of the grain boundary shown in Figure 13.8, it is apparent that the ln R term in the sum $E_S + E_I$ does not vanish since the grain boundary dislocations are of the same sign, so that the E_I term as well as E_S is positive. In this case R is chosen as L_o, that is, the range of the stress field associated with the grain boundary. It would thus appear that the law of the conservation of Burgers vectors does not hold for a grain boundary. This conclusion, however, is not true since our dislocation description of a grain boundary has thus far been incomplete. We must also add an additional vertical array of dislocations to each side of the boundary as shown in Figure 13.8. These arrays must be such that their Burgers vectors equal the negative of those in the grain boundary so that once again the law of conservation of Burgers vectors is obeyed. The computer solution of the grain boundary problem associated with Figure 13.8 shows that in an infinite crystal the rightmost and leftmost vertical dislocation arrays are spaced at a distance L_o from the boundary. These dislocations serve to screen the stress fields associated with the dislocations in the grain boundary and thus behave in a sense as surface dislocations. As the dimensions of the crystal decrease below L_o these surface dislocation arrays loose their uniformity and congregate closer about each of the arrays of crack arrays associated with the grain boundary. It is important to note that in the present model the ln R terms in Eqs. 13.19 and 13.20 cancel with one another. The grain boundary configurations shown in Figure 13.8 thus takes accurate account of the core size r_o and lattice size R in a manner that has been lacking in previous models. Finally, it should be emphasized that theories of grain boundaries, such as those espoused by Bollmann (1970) are essentially geometric models and say nothing with respect to elastic strain energy and surface energy.

13.3 SURFACE TENSION IN LIQUIDS AND SOLIDS

There has been a great deal of confusion in the literature with respect to the meaning of surface tension (Christian, 1965). In what follows, it is shown that all ambiguity associated with the problem vanishes upon incorporation of the concept of surface dislocation. Consider first the rectangular solid body shown in Figure 13.10*a*. It is clear that the surface of the body possesses a certain excess energy compared to the interior of the crystal. This arises from the fact that there are less atomic bonds associated with the surface atoms. In order to reduce this surface energy, the surface contracts. This in turn places the interior of the body in compression. The description of this state of internal stress can be described in terms of a continuous array of surface dislocations such as is depicted in Figure 13.10*a*. More specifically, as in the case of Figures 5.11*b* and 5.13*b*, we may view these surface dislocations as surface dislocation dipoles. One such dipole is drawn at the lower right of Figure 13.10*a*. The dotted portion of the dipole, in contrast to Figures 5.11*b* and 5.13*b*, no longer corresponds to the creation of new surface, but instead to the elimination of old surface.

Similar to Eq. 13.19, the total energy associated with the configuration shown in Figure 13.10*a* may be written as

$$E_T = E_S + E_I + E_\gamma \tag{13.24}$$

where E_S and E_I are the self and interaction energies associated with all of the surface dislocations. The sum of these terms is obviously positive since it corresponds to the strain energy within the body. On the other hand, E_γ corresponds to the energy associated with the decrease in the area of free surface and is obviously negative. Equilibrium is attained when the strain

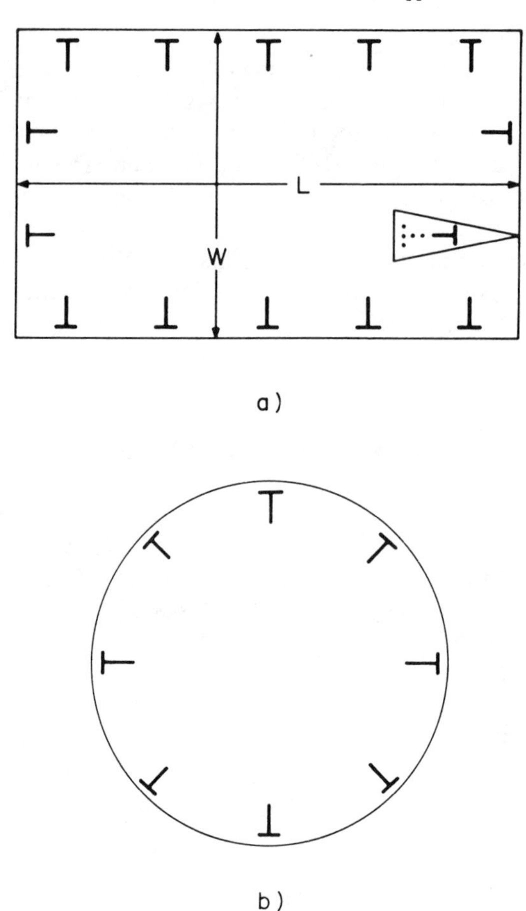

Figure 13.10 Surface tension of a *a*) solid and *b*) liquid described in terms of surface dislocations.

energy associated with the total number and distribution of surface dislocations is just balanced by the decrease in surface energy associated with these dislocations. Such calculations can easily be carried out numerically. It can also be shown (Jagannadham and Marcinkowski, 1978i) that as the ratio of L/W decreases to unity and the area kept constant, E_T given by Eq. 13.24 decreases to a minimum. A further decrease in energy can occur if the body changes its shape from a square to a circle such as shown in Figure 13.10*b*. In this case, the surface dislocations become uniformly distributed over the surface of the circle. It is also not difficult to see that Figure 13.10*b* represents precisely the shape taken by a liquid droplet so that the surface dislocation model of surface tension also applies to liquids as well as solids. It should be pointed out that a dislocation model for surface tension was first proposeed by Herring (1950), but not in terms of the present model of surface dislocations and was thus incomplete if also in fact somewhat incorrect. In the case of the configuration shown in Figure 13.10*b*, all shear components of stress within the body vanish, which in fact must be the case for a liquid which cannot support such stresses. Such, however, is not the case for the solid body depicted in Figure 13.10*a*, and such shear stresses in fact increase as the ratio L/W increases. The present results can also be applied to determine the shape of liquids in contact with solids, gases, or with liquids of a different kind (Jagannadham and Marcinkowski, 1978i). More generally, the surface dislocation model of liquids can be used to solve any boundary value problem associated with the behavior of liquids and thus provides a new and powerful insight into the disciplines of fluid mechanics and dynamics.

REVIEW

The theoretical foundations laid in the previous chapters have been applied to three broad classes of materials problems. First of all, the elastic shear crack shown in Figure 13.3*a* has been treated in detail. In particular, it has been shown that the stress-free boundary conditions required on both faces of the crack can be satisfied by the introduction of surface or crack dislocations on both faces of the crack. The total energy of such a crack can then be conveniently written as shown by Eq. 13.1. This of course is an elastic crack and vanishes when the applied stress is removed. On the other hand, such an elastic crack may lower its energy by allowing the crack dislocations to revert to quantized crystal lattice dislocations which leave the crack and move into the matrix against a lattice friction stress. Such a crack is depicted in Figure 13.3 where the dislocations shown dotted may be viewed as representing the plastic zone. The total energy of this plastic crack is given by Eq. 13.11. If now the external stress is removed from the plastic crack, it does not collapse, as was the case for an elastic crack, but instead becomes stabilized in the manner shown by Figure 13.3. The dislocations within the crack, however, now change their sign to become anti-crack dislocations. These anti-crack dislocations are simply surface dislocations which satisfy the stress-free boundary conditions arising from the dotted crystal lattice dislocations which are kept from running back into the crack by their lattice friction stress. This particular dislocation model of a crack allows any type of crack problem to be treated with a degree of both physical and mathematical insight that has hitherto been impossible. For example, the fatigue type crack can now be treated precisely.

The problem of grain boundaries can also now be formulated in great detail. Consider for example the grain boundary shown in Figure 13.8 which is a more detailed version of that given in Figure 7.5. It can be seen that the grain boundary can be viewed as a periodic arrangement of asymmetric cracks upon whose surfaces are distributed surface dislocations. The size of the cracks $L_o - L$ is determined by a balance between the elastic strain energy which trys to increase their length and the surface energy which trys to minimize their length. The total energy is given by Eq. 13.19. In order to complete the dislocation model of the grain boundary, it is necessary to add a set of screening dislocation arrays separated from the boundary by a distance R. The total Burgers vector of the screening array just balances that of the surface dislocations within the grain boundary and thus satisfies the very powerful conversation of Burgers vector law. These screening dislocations make it unnecessary to guess at the approximate value of crystal size R contained in Eqs. 13.21 and 13.22, since as Eq. 13.23 indicates, the energy of the arrays in Figure 13.8 become independent of R.

Finally, Figure 13.10 shows that the concept of surface tension is intimately tied up with that of surface dislocations. In particular, when an element of surface area decreases so as to reduce its energy, a surface dislocation is the immediate result. Equilibrium is established when the decrease in surface energy is just balanced by a corresponding increase in elastic strain energy due to the formation of surface dislocations. The various configurations of liquids in contact with solids, gases or other liquids can also be treated exactly by the methods of surface dislocations. This formulation may also be viewed as representing the initial stages in the construction of a dislocation theory for the behavior of liquids and would seem to justify the use of the word matter rather than solids in the title of this monograph. Proceeding a step further, it would also seem plausible that the behavior of gases could also be formulated in terms of dislocation theory, in turn leading to a truly unified theory of the mechanical behavior of matter. The extension of the concepts presented in the previous chapters seem to also be applicable to the solution of cosmological problems. For example, black holes which give rise to severe distortions in the space-time continuum, seem to be analogous to the effect produced by inclusions in matter. In fact, it has been recently shown (Bloomer, 1976) that torsion can be associated with black holes as well as with the singularities associated with the creation and collapse of the universe.

References

Anthony, K.-H., (1970a), *Fundamental Aspects of Dislocation Theory*, edited by J. A. Simmons, R. deWit, and R. Bullough, Vol. 1, National Bureau of Standards Special Publication 317, Washington, D.C. 20402, p. 637.

Anthony, K.-H., (1970b), *Archive for Rational Mechanics and Analysis*, **39**, 43.

Anthony, K.-H., (1971), *Archive for Rational Mechanics and Analysis*, **40**, 50.

Aris, R., (1962), *Vectors, Tensors and the Basic Equations of Fluid Mechanics*, Prentice-Hall, Inc., Englewood Cliffs, N.J.

Bloomer, I., (1976), "Modifications to General Relativity Induced by Torsion of Space-Time," Ph.D. Thesis, Queen Mary College, University of London.

Bollmann, W., (1970), *Crystal Defects and Crystalline Interfaces*, Springer-Verlag, New York.

Christian, J. W., (1965), *The Theory of Transformations in Metals and Alloys*, Pergamon Press, New York.

Coburn, N., (1970), *Vector and Tensor Analysis*, Dover Publications, Inc., New York.

deWit, R., (1960), *Solid State Physics*, Vol. 10, edited by F. Seitz and D. Turnbull, Academic Press, New York, p. 249.

deWit, R., (1973), *Journal of Research of the National Bureau of Standards—A. Physics and Chemistry*, **77**A, 49, 359, 607.

deWit, R., (1978), "Review of the Relation Between the Continuum Theory of Lattice Defects and Non-Euclidean Geometry in the Linear Approximation," article to be published.

Eringen, C. A., (1966), *Continuum Physics*, Vol. 1, Academic Press, New York.

Eshelby, J. D., (1956), *Solid State Physics*, Vol. 3, edited by F. Seitz and D. Turnbull, Academic Press, Inc., New York, p. 79.

Flügge, W., (1972), *Tensor Analysis and Continuum Mechanics*, Springer-Verlag, New York.

Frank, F. C., (1951), *Philosophical Magazine*, **42**, 809.

Freudenthal, A. M., (1958), *Handbuch der Physik*, edited by S. Flügge, Vol. VI, Springer-Verlag, Berlin, p. 229.

Fung, Y. C., (1965), *Foundations of Solid Mechanics*, Prentice-Hall, Inc., New Jersey.

Gel'Fond, I. M. and G. E. Shilov, (1964), *Generalized Functions*, Vol. 1, *Properties and Operators*, Academic Press, New York.

Gołąb, S., (1974), *Tensor Calculus*, Elsevier Scientific Publishing Company, Amsterdam.

Head, A. K., (1953), *The Philosophical Magazine*, **49**, 92.

Herring, C., (1950), *The Physics of Powder Metallurgy*, edited by W. E. Kingston, McGraw-Hill, New York.

Hill, R., (1950), *The Mathematical Theory of Plasticity*, Oxford University Press, London.

Hirth, J. P. and J. Lothe, (1968), *Theory of Dislocations*, McGraw-Hill Book Company, New York.

Hoffman, O. and G. Sachs, (1953), *Introduction to the Theory of Plasticity For Engineers*, McGraw-Hill Book Company, Inc. New York.
Jagannadham, K. and M. J. Marcinkowski, (1978a), *International Journal of Fracture*, **14**, 155.
Jagannadham, K. and M. J. Marcinkowski, (1978b), *Physica Status Solidi*, **a50**, 293.
Jagannadham, K. and M. J. Marcinkowski, (1978c), *Journal of Materials Science and Engineering*, in publication.
Jagannadham, K. and M. J. Marcinkowski, (1978d), *Journal of Materials Science and Engineering*, in publication.
Jagannadham, K. and M. J. Marcinkowski, (1978e), *International Journal of Fracture*, in publication.
Jagannadham, K. and M. J. Marcinkowski, (1978f), *International Journal of Fracture*, submitted for possible publication.
Jagannadham, K. and M. J. Marcinkowski, (1978g), *International Journal of Fracture*, submitted for possible publication.
Jagannadham, K. and M. J. Marcinkowski, (1978h), *Physica Status Solidi*, **a50**.
Jagannadham, K. and M. J. Marcinkowski, (1978i), *Journal of Materials Science*, in publication.
Kondo, K., (1955), "Memoirs of the Unifying Study of the Basic Problems in Engineering Sciences by Means of Geometry," *Gakujutsu Bunken* Fukyu-Kai, Tokyo, p. 458.
Kröner, E., (1958), *Kontinuumstheorie der Versetzungen und Eigenspannungen*, Springer-Verlag, Berlin.
Kröner, E., (1959), *Archive for Rational Mechanics and Analysis*, **4**, 273.
Kröner, E., (1964), *Vorlesungen über Theoretische Physik*, von Arnold Sommerfeld, Vol. II, Akademische Verlagsgesellschaft, Geest & Portig, Leipsig.
Kröner, E., (1966), *Theory of Crystal Defects*, edited by B. Gruber, Academic Press, New York, p. 231.
Kunin, I. A., (1965), "Methods of Tensor Analysis in the Theory of Dislocations," A supplement to J. A. Schouten's *Tensor Analysis for Physicists*, Moscow. Also available in English translation from the U.S. Department of Commerce, Springfield, VA 22151.
Lardner, R. W., (1974), *Mathematical Theory of Dislocations and Fracture*, University of Toronto Press, Toronto.
Likachev, V. A. and R. V. Kaerov, (1975), *Introduction to the Theory of Disclinations*, Leningrad University, Leningrad.
Marcinkowski, M. J., (1970), *Fundamental Aspects of Dislocation Theory*, NBS Special Publication 317, Vol. I, edited by J. A. Simmons, R. deWit and R. Bullough, p. 531.
Marcinkowski, M. J., (1971), *Advances in Materials Research*, edited by H. Herman, **5**, 443, Wiley-Interscience, New York.
Marcinkowski, M. J., (1972), *Electron Microscopy and Structure of Materials*, edited by G. Thomas, R. M. Fulrath and R. M. Fisher, The University of California Press, Berkeley, CA, p. 382.
Marcinkowski, M. J., (1975), *Journal of Applied Physics*, **46**, 496.
Marcinkowski, M. J., (1976), *Physica Status Solidi*, **a38**, 223.
Marcinkowski, M. J., (1977a), Final Technical Report, "Study of the Structure and Deformation of Internal Boundaries," Grant No. DMR-7202944, National Science Foundation, Washington, D.C.
Marcinkowski, M. J., (1977b), *Acta Crystallographica*, **A33**, 865.
Marcinkowski, M. J., (1977c), *Crystal Lattice Defects*, **7**, 131.
Marcinkowski, M. J., (1977d), *Archives of Applied Mechanics*, **29**, 313.
Marcinkowski, M. J., (1977e), *Philosophical Magazine*, **36**, 1499.
Marcinkowski, M. J., (1978a), *Crystal Lattice Defects*, submitted for possible publication.
Marcinkowski, M. J., (1978b), *Physica Status Solidi*, submitted for possible publication.

Marcinkowski, M. J., (1978c), *Acta Mechanica*, in publication.
Marcinkowski, M. J., (1978d), *Acta Technica*, submitted for possible publication.
Marcinkowski, M. J., (1978e), *Acta Mechanica*, in publication.
Marcinkowski, M. J., and E. S. P. Das, (1971), *Physica Status Solidi*, **a8**, 249.
Marcinkowski, M. J., and E. S. P. Das, (1974), *International Journal of Fracture*, **10**, 181.
Marcinkowski, M. J., and Dwarakadasa, (1973), *Physica Status Solidi*, **a19**, 597.
Marcinkowski, M. J., and K. Sadananda, (1973), *Physica Status Solidi*, **a18**, 361.
Marcinkowski, M. J., and K. Sadananda, (1975), *Acta Crystallographica*, **A31**, 280.
Marcinkowski, M. J., and K. Sadananda, (1977) *Acta Mechanica*, **28**, 159.
Marcinkowski, M. J., K. Sadananda, and W. F. Tseng, (1973), *Physica Status Solidi*, **a17**, (1973).
Nabarro, F. R. N., (1967), *Theory of Crystal Dislocations*, Oxford at the Clarendon Press, Oxford.
Nye, J. F., (1957), *Physical Properties of Crystals*, Oxford at the Clarendon Press, Oxford.
Orowan, E., (1934), *Z. Phys.* **89**, 634.
Polanyi, M., (1934), *Z. Phys.* **89**, 660.
Prager, W., (1959), *An Introduction of Plasticity*, Addison-Wesley, London.
Sadananda, K., (1976), International Conference on Computer Simulation for Materials Application, Gaithersburg, MD.
Sadananda, K., K. Jagannadham, and M. J. Marcinkowski, (1978), *Physica Status Solidi*, **a44**, 633.
Sadananda, K. and M. J. Marcinkowski, (1972), *Journal of Applied Physics*, **32**, 2609.
Schouten, J. A., (1951), *Tensor Analysis for Physicists*, Oxford at the Clarendon Press.
Schouten, J. A., (1954), *Ricci-Calculus*, Springer-Verlag, Berlin.
Sneddon, I. N. and D. S. Berry, (1958), *Handbuch der Physik*, edited by S. Flügge, Vol. VI, Springer-Verlag, Berlin, p. 1.
Sokolnikoff, I. S., (1964), *Tensor Analysis* 2nd Ed., John Wiley & Sons, Inc., New York.
Spain, B., (1953), *Tensor Calculus*, Interscience Publishers, Inc., New York.
Taylor, G. I., (1934), *Proc. R. Soc.*, **A145**, 369.
Truesdell, C. and R. A. Toupin, (1960), *Handbuch der Physik*, edited by S. Flügge, Vol. III/1, Springer-Verlag, Berlin, p. 226.
Volterra, V., (1907), *Annls. Scient. Éc. Norm. Sup.*, Paris, **24**, 401.
Wayman, C. M., (1964), *Introduction to the Crystallography of Martensite Transformations*, The MacMillan Company, New York.
Zorawski, M., (1967), *Théorie Mathématique des Dislocations*, Dunod, Paris.

Index